MOUNTAIN BUILDING GEOLOGY
IN THE
PACIFIC NORTHWEST

NED BROWN

Headed to the north face of Mt Shuksan. Doug McKeever & Al Doan 1971

ISBN 9780999527801

LCCN 2017957740

Printed in Bellingham WA, USA.

Village Books
1200 11th Street
Bellingham WA 98225
360 671-2626

1st edition.

CONTENTS

Cover: The author on Magic Mountain, North Cascades. Sonja Nelson photo.

Alpine meadow on Shuksan Arm

INTRODUCTION

Mountain belts around the globe store a record of continental growth by tectonic plate collisions and additions of molten rock (magmatism), and consequently are of great interest in geologic exploration. Considerable geologic theory has come, for example, from studies of the European Alps, Himalaya, and Andes. The mountains of interest here are in the same league. The North Cascades and southern Coast Mountains of British Columbia are a part of what we call the North American Cordilleran Orogen—a mountain belt extending from Alaska to Mexico. This great range has much to tell about how the western margin of the North American continent has grown oceanward hundreds of miles in the last 250 million years of earth history.

I came to the Pacific Northwest in 1966, having grown up on the flatlands of Minnesota, to take a job at Western Washington State College (now Western Washington University), at the edge of the North Cascades. I had explored the Cascades on climbing vacations away from graduate school at Berkeley. Given choices, this was a no-brainer. Don Easterbrook offered me the job over the phone and I accepted sight unseen. My expectations panned out. Much intriguing geology and mountain adventure have come my way.

My work in the mountains of the Pacific Northwest began at the doorstep of WWU in the local Cascades foothills in the late 1960s. Following the rocks, and enlisting graduate students, the studies

High camp on the west shoulder of Mt. Urquhart in the southern Coast Mountains of British Columbia.

5

extended across the North Cascades through the 1970s and into the 80s. Then, looking around for more clues that could solve the complicated problems of Cascades origins, students and I traced the geology north across the border into the British Columbia Coast Mountains in the late 1980s through the 90s. Subsequently, with body parts pretty much spent, I retired from the mountains and turned my attention to the more accessible San Juan Islands.

Of course, the students and I were not geologic pioneers in the Cascades. Geologic mapping by Peter Misch and his students at the University of Washington in the 1950s and early 60s accomplished a major achievement in establishing a structural framework for the Cascades that has largely been substantiated by subsequent workers. Misch, born in Germany, worked in the Alps, Pyrenees, and Himalayas before he emigrated to the United States, bringing much experience in mountain geology.

Subsequent to Misch, there have been many other geologists. Particularly notable are the contributions of Rowland Tabor and Ralph Haugerud, of the U.S. Geological Survey, who systematically mapped in detail virtually all of the North Cascades. They produced elegant maps at scales of both 1:100,000 and 1:200,000, plus numerous accompanying USGS reports with extensive rock descriptions. Tabor and Haugerud have also published, through The Mountaineers, an excellent field guide written for the layman. Numerous other geologists have in recent years published detailed aspects of certain parts of the Cascades in scientific journals. Bob Miller, of San Jose State University, with colleagues, has published considerably on plutonism and tectonics of the Cascade Core, including technical field guides. All these materials are drawn on extensively in this book and are referenced in the last pages.

In our British Columbia study area, much geology was known by the time we got there, particularly through work of Jim Roddick and Jim Monger of the Geological Survey of Canada, and also a number of student theses from the University of British Columbia and the University of Washington.

Many hardy students were attracted into my sphere with the prospect of geologic research in the mountains. Some 25 M.S. students completed theses, and many more were undergraduate field assistants. Of course, most of the mapping was in the woods, in some places perilously steep and slippery, clogged with Devils Club, and defended by insects. Somehow in this environment we found fun—rating bush-whacking moves on a rock climbing scale, calling out to advertise our presence to bears, dipping in a pool, having a PBR back at camp. The geology told in this book derives from the combined effort of the many involved.

Paul Furlong looking for a way through the Devil's Club.

A large part of the study was supported by grants from the National Science Foundation, applications for which require a statement of how the general public will gain value from the proposed study. So, this book is in part an effort to bring our findings to a broad audience. The grants were vital — providing summer salaries, buying food, renting vehicles, and paying for the very important helicopter support.

A goal of our research, and a purpose of this book, is to gain an understanding of what has happened at depth in the earth to create this mountain range. Mapping of the rock units and the observed faults and folds across the landscape goes a long way to this end. But much can be gained also from radiometric dating of the rocks, determination of the depth of formation of the rocks from their mineralogy, and documenting movement directions of the rocks evidenced by their structural fabrics. A full exposition of these details goes beyond the limits of my intended audience here. (It took me five years of graduate school to learn this stuff.) In this text I will present in at least some depth the concepts of "geochronology", "thermobarometry", and rock "kinematics" to support interpretations. I urge the reader to jump over any amount of this explanation and documentation if it is burdensome, and move on to the general findings and conclusions. The Glossary, beginning on page 139, will help.

Rock exposures for the casual geologist are not so easy to come by in the Cascade and Coast Mountains because, up to about 4000 feet or so in elevation, outcrops are commonly covered by moss and forest; and in the alpine zone where exposures are excellent, getting there is problematic. With these difficulties in mind, this book presents for the reader who may not be a mountaineer many photos of rocks at remote localities important to the overall goal of understanding how mountain building works. In addition, views of the rocks through the microscope are used abundantly for presenting the same kind of critical evidence on a different scale.

Theories explaining the origin of the North Cascades and co-extensive southeastern Coast Mountains of British Columbia are almost as numerous as the scientists who have worked there. The mental gymnastics involved in balancing and evaluating all the different evidences spins the brain, leads to controversy, and gives meaning to life (for some anyway). Part of the presentation here is to bring forth these discussions. Some of my friends and colleagues would tell this story differently.

Within the chapters there is some guidance on places to see the geology; road access and good localities are marked on maps. Off-trail and alpine areas of interest to climbers and other hardy types are also noted in the text. Field guides for this region published by others are noted in the *references* section of this book.

View south from Mount MacFarlane in southern British Columbia. From left to right: Mt. Shuksan, the Pleiades, Mt. Larrabee, American Border Peak.

A brief introduction to the geology and the regional layout is worthwhile here. The map below shows the main geologic rock elements, and the area of focus for the book. The Coast Plutonic Complex (CPC) in British Columbia, and its extension into Washington as the Cascade Core, constitutes a mixture of granitic plutons and the sedimentary and volcanic "country rock" that the plutons intrude. The Coast Plutonic Complex is a magmatic arc, extending some 1000 miles northward to Alaska. The country rock is mostly far-travelled rock, moved by plate tectonics and accreted to the continental margin. Such rock, exotic in origin to the continent, we term a terrane. Other nearby but far-traveled terranes are *Wrangellia*, making up Vancouver Island, and *Quesnellia*, east of the arc and comprising the bedrock of much of southern British Columbia.

A different set of rocks makes up the San Juan Islands - northwest Cascades thrust system (SJI-NWC), and extends partway into the B.C. Coast Mountains. This complex is a stack of thrust sheets that apparently has had little to do with the Coast Plutonic Complex arc, other than being thrust over the south end. Rocks of the individual thrust sheets are also terranes, mostly of oceanic origin. They accreted to the continent somewhere to the south, were carried to great depth ("subducted"), then later moved into their present place—all by plate tectonics.

The terranes in both the CPC and the SJI-NWC have been affected by metamorphism; this is recrystallization at depth due to increased pressure and temperature. We can determine the depth from mineral compositions and learn much about how the mountains of this region developed.

To get everyone up to speed on the science of tectonic processes, metamorphic mineralogy, and how rock ages are measured, the first few chapters present the basic elements of these tools. Let's get started digging into what we know and how geologists figure it out, as well as where uncertainties point to the need for further work.

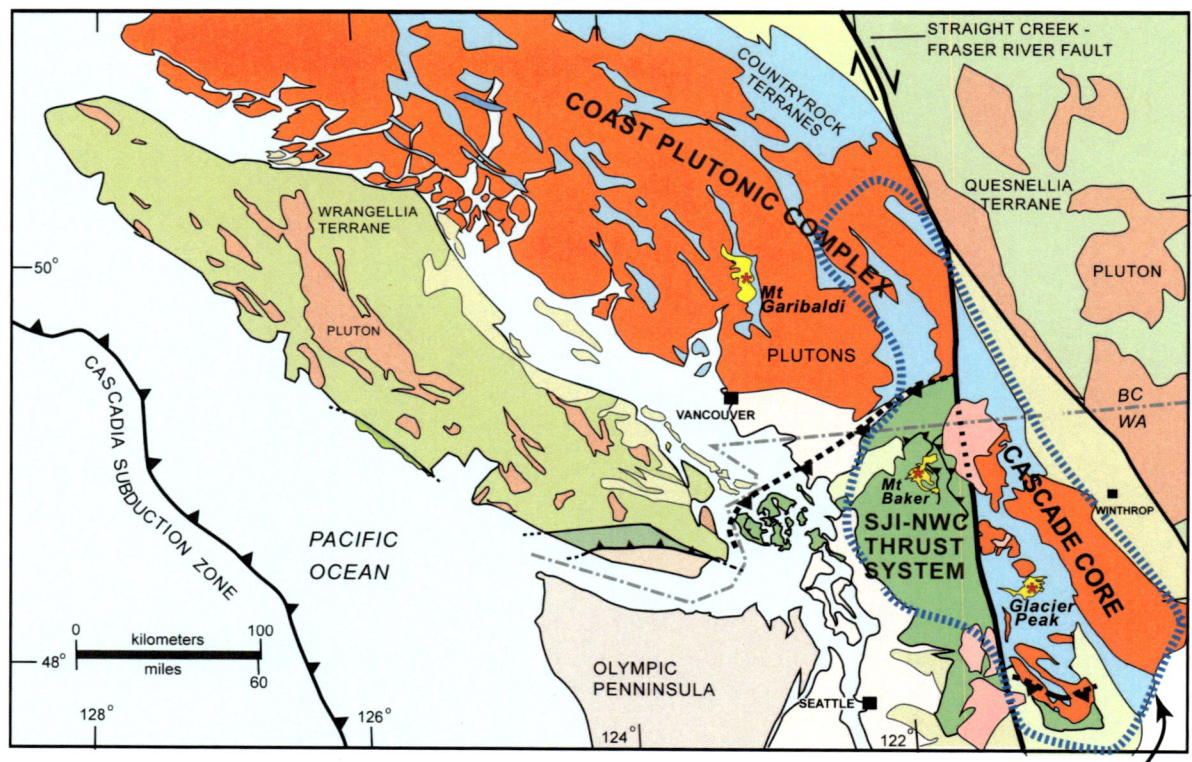

MOUNTAIN GEOLOGY
DESCRIBED IN THIS BOOK

ACKNOWLEDGEMENTS

This book summarizes a lifetime of geologic investigation into the origins of mountains lying within a few hours drive from my employment at Western Washington University and my home in Bellingham. Countless geologic colleagues, students, and friends have contributed to this enterprise. Master's degree students and undergraduate assistants shouldered much of the heavy work of back-country mapping.

Various universities, over the decades, generously gave me access to their microprobe facilities for mineral analyses: University of Washington, Cambridge University, Kyoto University, University of Bern, and Otago University. Zircon age-dating for my projects has been done by Nick Walker, Bill McClelland, and recently by myself with training and help from George Gehrels and staff in the LaserChron Center at the University of Arizona.

My understanding of, and speculation about, the local mountain geology has evolved over the years in large part from many constructive field trips, office discussions, and emails with persons of like interest: Peter Misch, Rowland Tabor, Ralph Haugerud, Bob Miller, Jim Talbot, Liz Schermer, and Jim Monger. But certainly any of this group would question some interpretations made here.

Critical funding for the research came from the National Science Foundation, supporting field and laboratory studies, for both me and students.

Reviewers of the book took on a lot of work. I am most grateful. For editing, I thank my wife Linda and sister-in-law Ellen Brown. Beyond the usual grammar, punctuation, etc., these two warriors tried to make sense of the text and rein in the "geo-speak"—Ellen's manuscript draft is much marked up with "GS". Linda Earl, a retired professional editor and my next-door neighbor, fine-tuned the manuscript.

Geologic reviewers were: Jim Monger, who looked at the whole manuscript; Bob Miller, who focused on the Coast Plutonic Complex; Sue DeBari, who checked my explanations of the plumbing of arc processes; and Harold Stowell, who advised on description of the Swakane Gneiss.

Finally, I thank Brendan Clark of Village Books for guidance in formatting the manuscript and for his construction of the book cover.

Kangaroo Ridge, from Washington Pass overlook

Fall colors at Table Mountain

PART I — BACKGROUND GEOLOGY

CHAPTER 1

PLATE TECTONICS

The North American continent has grown out greatly on its western flank in the last few hundred million years. Virtually all of this growth happened by way of plate tectonics. Thus, we start with the basic concepts of this global phenomenon, and in later chapters consider how it applies to mountain building in the Pacific Northwest.

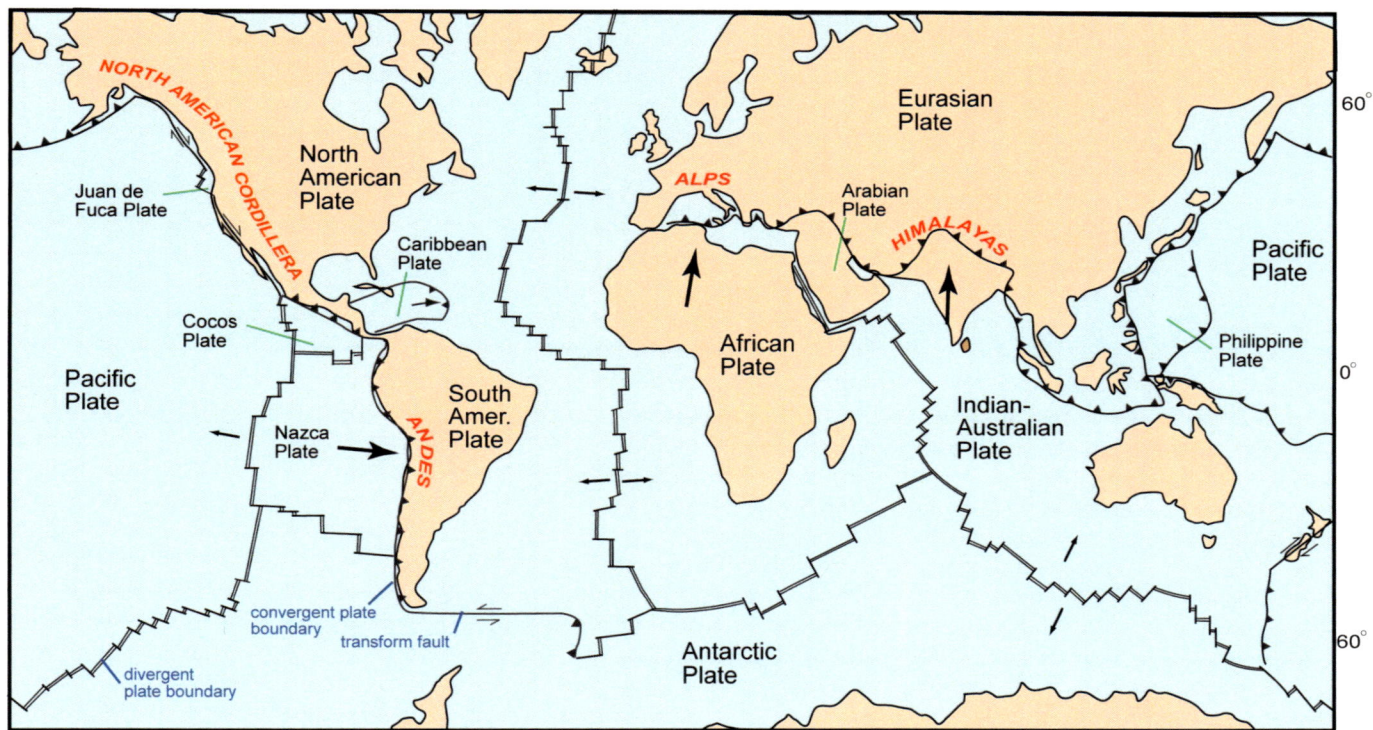

Fig. 1-1 Earth's tectonic plates. Plates pull apart at ocean ridge systems, slide sideways relative to one another along transform faults, and at some boundaries converge, creating mountains and magmatic arcs.

Looking at this map of continents and oceans (Fig. 1-1), though without the recently added black lines, German meteorologist Alfred Wegener, in the early 20th century, proposed that the continents were once all glommed together and somehow moved apart into their present array. "Continental drift"— a crazy idea? He cited the apparent "fit" of continents on opposite sides of the Atlantic if the ocean were closed. And he noted a match of fossil plants and geology from South America to Africa. This hypothesis languished, as no mechanism was known for continents to break up and wander about. At that time we knew so little about what goes on at depth in the earth.

A wealth of information emerged in the late 1940s to early 1960s as researchers documented earthquakes passing through the deep earth, mapped the ocean floor by sonar and deep-sea dredging, and drilled kilometers into the ocean crust (the "Mohole" project). From these studies we learned that the earth's sphere is layered. The outer layer is a relatively thin crust. Ocean crust is 5 to 10 miles thick and made of relatively young (mostly less than 150 million years old) basalt with a thin coating of sediments.

Continental crust is 20 to 30 miles thick, and the rocks are pretty much as we see at the surface—granite, schist, limestone, etc. In places the rocks are stunningly old—broad continental areas are 2-4 billion years old.

Under the crust is an 1800 mile thick layer, termed the mantle, consisting of Mg-Fe silicate minerals—olivine-pyroxene rock in the upper levels. We can actually put our hands on this rock where fragments have been carried up in volcanoes, or as slabs shoved up along deep faults (the Twin Sisters range in the western Cascades, Chapter 7).

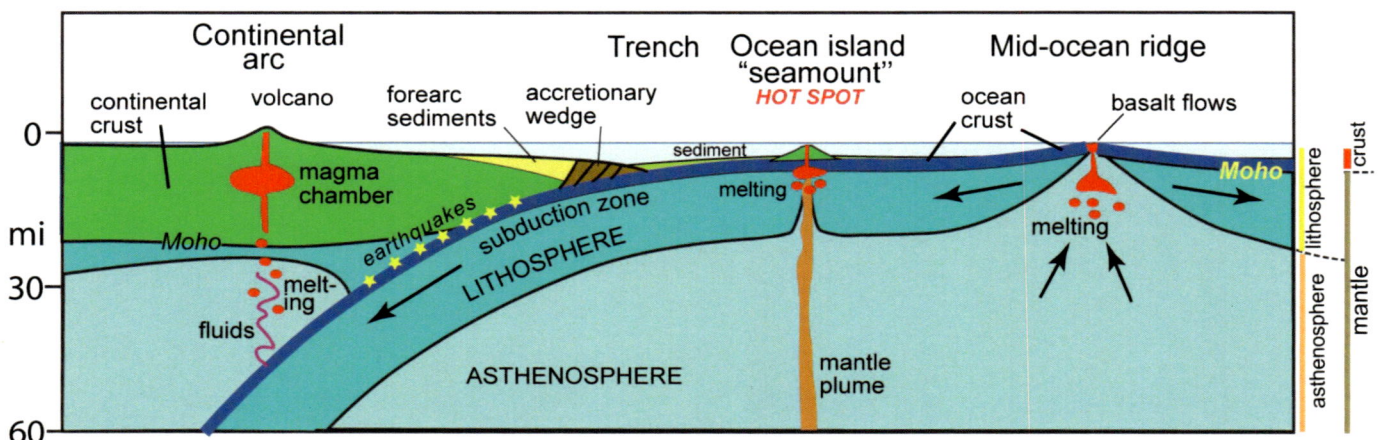

Fig. 1-2 Schematic cross section showing tectonic layers in the outer earth and mobility of plates: the "lithosphere" sliding on the "asthenosphere".

The crust-mantle boundary, termed the Mohorovicic discontinuity ("Moho" of Fig. 1-2) after its discoverer, is distinctly marked by a jump in earthquake wave velocity, so is well documented around the globe. Very important to understanding "continental drift" is the presence in the mantle of a "low velocity zone" of earthquake waves extending from about 50 to 200 miles depth, termed the asthenosphere (Fig. 1-2). These slow earthquake waves are the "shear waves" that shake the rock sideways. The slowing down of the shear waves reflects weakness of shear strength of the asthenosphere; the rock is solid but exhibits plasticity, like clay. The overlying upper mantle and crust can be pushed down into the asthenosphere, and lateral forces can actually slide the upper mantle, crust, and continents across the globe (continental drift!).

At depth below the mantle is the earth's core, some 1500 miles in

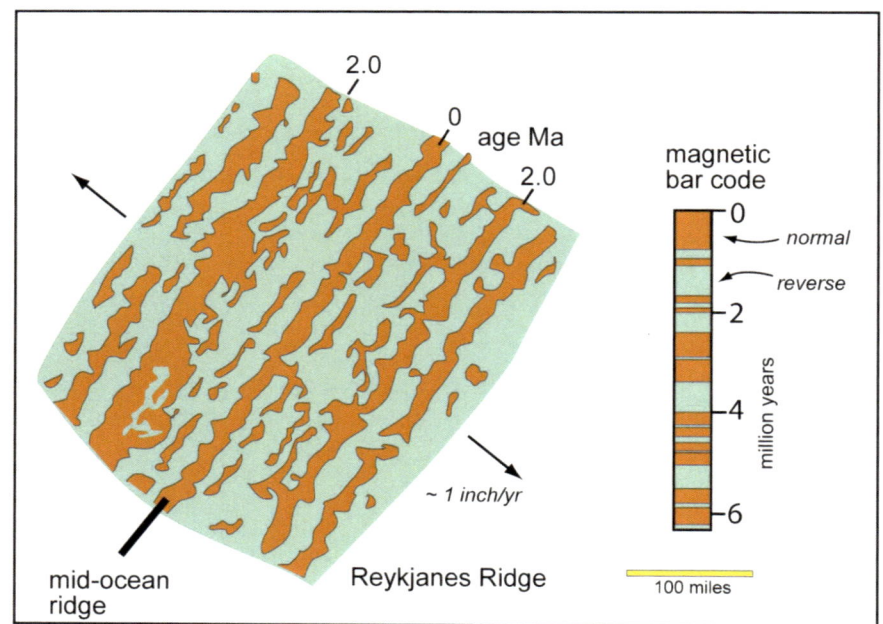

Fig. 1-3 Patterns of magmatic intensity measured along the mid-ocean ridge system in the North Atlantic.

diameter. The core is interpreted to be made of nickel and iron based on density, earthquake wave velocity, and the abundance of metallic nickel-iron meteorites. The core is partitioned into a solid inner part and a molten outer part. Earth's magnetic field, important to the analysis of "continental drift" as we shall see, apparently originates from electric currents generated by molten flow in the outer core.

Remnant Magnetism

Sonar mapping of ocean basins revealed, by the 1950s, very extensive deep sea topographic ridges—a puzzle. Concurrently, remnant magnetism in the sea floor rocks was discovered by magnetometers towed behind ships during the war looking for submarines. Later in the '50s, the same types of surveys were expanded to understand sea floor geology (Figs. 1-3, 1-4). The earth's magnetic field of the time was captured and preserved in the mineral "magnetite" during lava crystallization. Belts of stronger and weaker magnetism parallel to the ridges were discovered, posing another puzzle.

In the early 1960s, researchers recognized that the sea floor magnetic belts mark periods of normal and reverse magnetization: a flipping of the magnetic direction from north to south and back, somehow related to action in the dynamo in the earth's core that reverses the spin direction of currents of electrons. The same patterns of normal and reverse magnetization were found on land, and there the age of the magnetization was determined from isotopic dating of minerals in the rock (potassium-argon, uranium-

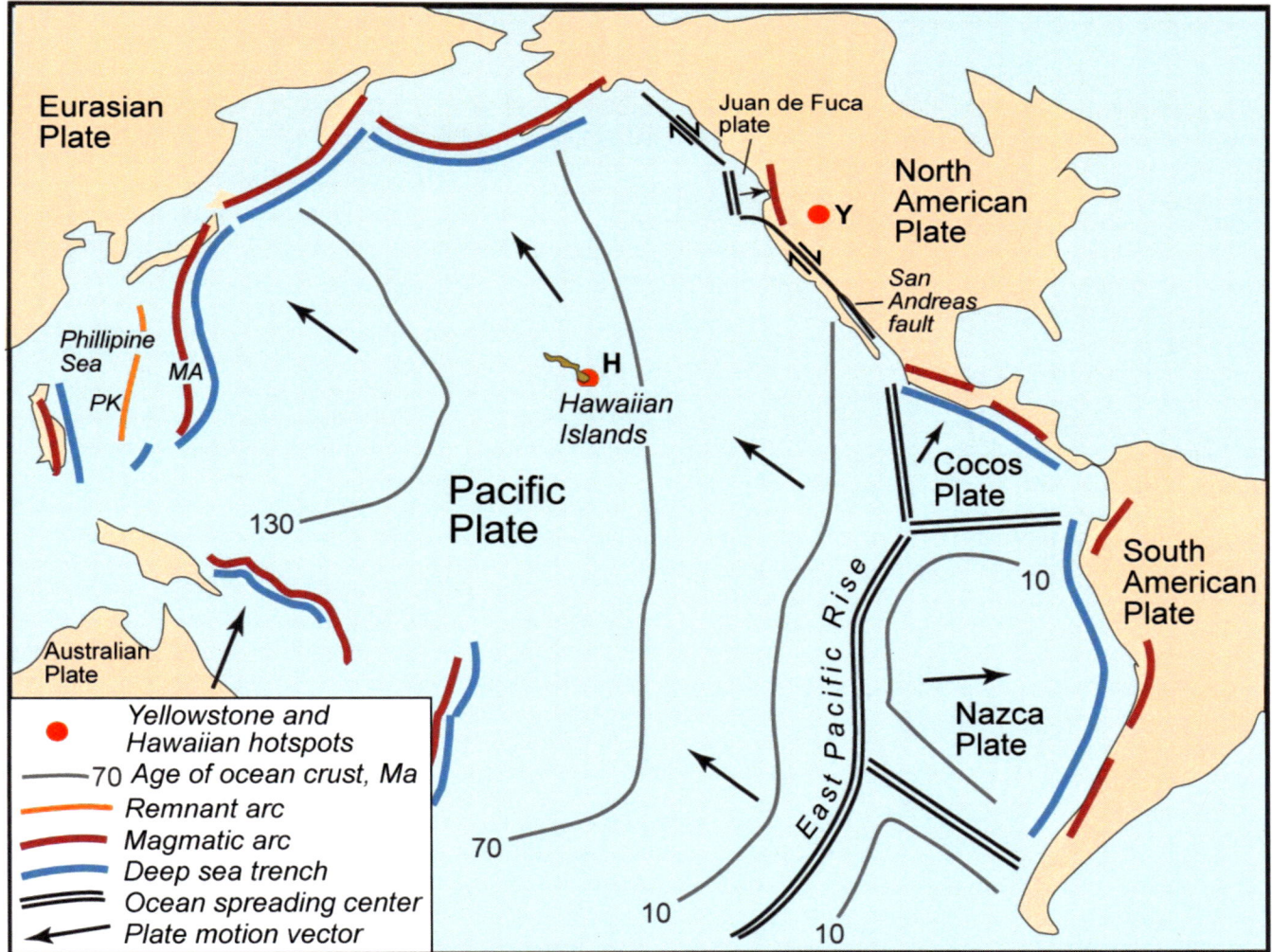

Fig. 1-4. Sea floor ages and plate motion directions in the Pacific basin.

lead, Chapter 3). From this a "geomagnetic polarity time scale" was established based on the distinctive patterns of normal and reverse magnetization—a tool analogous to tree ring chronology. The bedrock of ocean basins is virtually everywhere dated by the pattern of magnetic belts (Fig.1-4). Interestingly, the oldest sea floor rocks are only about 170 million years old, whereas continents in places are 3-4 billions of years old.

Plates

In the early 1960s, a number of geologists put all these curious features together to develop the "sea floor spreading" model. Along the ocean ridges is a central valley in which basalt lava erupts (Fig. 1-2). Accompanying the basalt eruptions, the ridges pull apart by rifting normal to the ridge axis. This action leads to growth of ocean crust at the ridges, and broadening of the ocean basin away from the ridges. Supporting the ridge system dynamics, fossils in ocean floor sediments show that ocean crust is systematically older moving away from the ridges.

So, following the geology of ocean ridge systems, we see that oceans expand. Here is a possible explanation for the drifting apart of South America from Africa, and North America from Europe, as the Atlantic Ocean expanded. A theory of global earth expansion (like inflating a balloon) had its day back in the 1950s. A fatal headwind was the significant problem of explaining where the added earth mass came from.

Earthquake studies in the 1940s and 50s solved this question of the "expanding earth". Independently, American and Japanese geophysicists (Benioff and Wadati) discovered that, along Pacific Ocean margins, earthquakes are distributed along planes that extend hundreds of kilometers down—dipping away from the ocean under South America on the east side and under volcanic island chains on the west side (Fig. 1-4). Earthquakes are caused by faulting, and along these earthquake zones the displacement is "dip slip"—up and down, not sideways as in "strike-slip". The motion shows the Pacific crust to be descending under the overlying continent and volcanic chain. This finding relieves the problem of the expanding earth. Ocean crust is generated at the ridges, but it is consumed at ocean margins by underthrusting (Fig. 1-2), in what are now termed subduction zones.

The globe is covered by plates —altogether some 12 in number. Bounding structures are spreading centers at ocean ridges and subduction zones, as we have considered. But there is also another type of boundary where plates slip sideways relative to each other, termed transform faults. Perhaps the most famous of these is the San Andreas Fault (Fig. 1-4).

Subduction and Accretion

At the convergent margin we find, commonly but not everywhere, deep sea trenches. The down-going ocean slab pulls the topography down with it. Some distance inland from the trench, great volumes of magma rise upward in elongate belts, generated by release of fluids in the subducting slab and melting in the overlying mantle and crust. These igneous belts, termed magmatic arcs, are a dominant feature of ancient western North American geology and the foundation of its mountains.

Also along the convergent margin, the incoming ocean crust that disappears down the subduction zone can deliver masses of rock material onto the edge of the continent. This accumulation of material, termed an accretionary wedge, is in places ocean crust, seamounts, old arcs, pieces of other distant continents, and sediment eroded from the continental margin. Much of this added material is carried down some distance on the down-going slab in a subduction zone (Fig. 1-2) before breaking off to be "under-plated" on the over-riding plate. Such rocks are metamorphosed—changed to high-pressure minerals, and are severely strained (smeared).

Hot Spots and "Polar Wander"

Seamounts, though a very minor part of ocean crust, play a major part in unravelling plate motions. These oceanic volcanoes form over a plume of very hot, plastic mantle peridotite reckoned to rise from near the mantle base. As the plume nears the surface, the peridotite partially melts owing to decreasing pressure—yielding basalt magma. The basalt intrudes upward to form an oceanic volcano, referred to as an "ocean island", or "seamount" where the mass is under water (Fig. 1-2). The rock chemical composition is distinctive of this origin. The Hawaiian Islands are a familiar example.

Ocean islands and seamounts form by an intersection of earth forces: the deeply anchored stationary mantle plume and the mobile ocean crust. In Hawaii we see a string of islands that are progressively older to the northwest, and extend below sea level as a chain of seamounts all the way to the Aleutian Islands, where they are subducted. The Big Island has the only active volcano. Putting these features together, and figuring that the mantle plume is fixed in position, we can measure the drift speed and direction of the Pacific plate over the Hawaiian mantle plume by determining ages for the string of ancient volcanoes, including the seamounts, as shown in Fig. 1-4. Such plumes, and ancient record of plumes in the form of chains of extinct volcanoes, occur in many places on the earth, both under oceanic and continental crust. They are called "Hot Spots".

Another tool is available in establishing plate motions. Above, we looked at how the remnant magnetism in sea-floor rocks dates the ocean crust. Another valuable aspect of remnant magnetism is that the magnetic vector, like a compass needle, points to the North (or South) Pole. Of course the magnetic pole moves about on a short-term basis, but averages in a set of related rocks to be at the earth's rotational pole.

The magnetic vector has an inclination related to its latitude of formation—horizontal at the equator, progressively more steeply inclined going toward the pole. Thus, paleomagnetic data for a rock literally point to the rotational pole of the time. Any departure of the ancestral pole for a rock from the current pole is referred to as "polar wander", but the data actually represent rock wander, i.e. plate drift.

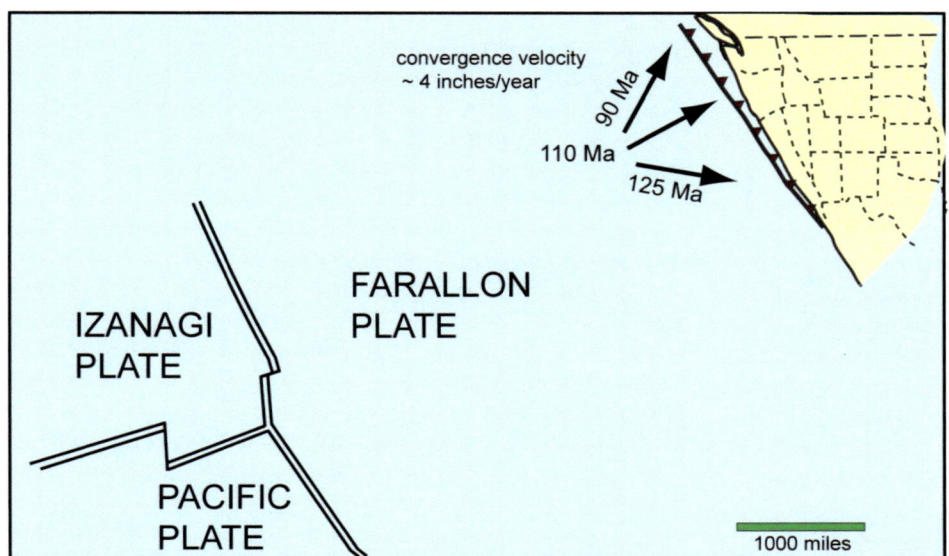

Fig. 1-5. Plate reconstruction for the Pacific Ocean basin and plate convergence with North America approximately 100 million years ago, from Engebretson et al. 1985. Reconstruction of the continental margin at 100 Ma is estimated by Severinghaus and Atwater, 1990.

The Farallon Plate

The geology of the Cascades and B.C. Coast Mountains is very much tied to the incoming oceanic plate. The mountain-building process we see on the Pacific Northwest continental margin owes its origin to collision, subduction, volcanism, and plutonism, all caused by effects of a converging oceanic plate. What did this plate interaction look like back in the day of orogeny (mountain building), 100 to 80 million years ago? We need to know the plate directions and velocities. Tools for this enterprise include: 1) hot spot tracks that show individual plate motions relative to the fixed underlying mantle, 2) magnetic stripes on the sea floor that mark and date ancient ridge positions, and link motions of plates on opposite side of the ridge, and 3) the "polar wander" of a plate indicated by the direction and inclination of the paleomagnetic vector.

My friend and colleague at Western Washington University, Dave Engebretson, took on the project of determining ancient plate motions for his PhD at Stanford. His solutions, published in 1985, are the current definitive model.

The geometry of describing a system of plates moving about on a globe, colliding, moving apart, and sliding past one another, boggles the mind. Dave was an undergraduate math major —that helps when it comes to unravelling the spherical geometry involved.

To start, we consider the present plate configuration of the Pacific (Fig. 1-4). Most of the Pacific Ocean is underlain by the giant Pacific Plate. Spreading ridges are present, but the plates on the opposite side from the Pacific Plate are small: Juan de Fuca in the northern Pacific, and Cocos and Nazca in the southern Pacific. Following plate tectonic theory, a large plate equal in size to the Pacific plate was generated at the ridge system; we call it the Farallon Plate (Fig. 1-5). Where did it go? Subducted under North and South America, Duh! The small plates are remnants. Where the Pacific Plate is directly against the North American Plate in California and British Columbia, subduction has entirely consumed the east-going plate (Farallon) and the spreading ridge; the contact is now a sideways moving fault—the San Andreas Fault. This is a transform fault.

To get a vision of plate geometry when the Cascade and B.C. mountains were developing ~100 million years ago, we need to back the Farallon Plate out of the subduction zone. A key element here is the record of plate formation marked by magnetic stripes on the Pacific Plate, which reflect the growth of the Farallon Plate. The Farallon Plate is considered to be a mirror image of the Pacific Plate, giving us the size and motion of the Farallon plate relative to the Pacific plate. This relationship, together with the hot spot movement record of the North American and ocean plates, yields direction and rate of motion in the collision zone between North America and the incoming Farallon Plate—-very helpful in understanding mountain building in the Pacific Northwest. Most times convergence was oblique, with an early coastwise north component and a later coastwise south displacement (Fig. 1-5).

Coast Plutonic Complex Continental Arc
Now let's look at the connection of the British Columbia Coast Mountains and related parts of the Cascades to subduction of the Farallon Plate. Rocks in these regions constitute a "continental arc" (Figs. 1-2, 1-4). Here we delve into the topic of the magma plumbing. For starters, we know from the global map of plates and subduction zones that the arcs form over a subducted plate; also, from the angle of subduction and the position of the arc, that the depth from the volcano to the magma source is fairly uniform at about 60 miles or so.

Where does the magma come from? One would think perhaps that the subducting plate simply melts in that zone below the arc. But, not the case—the plate remains essentially solid. The story begins at the oceanic spreading center (Fig. 1-6) where upwelling mantle peridotite (olivine + pyroxene rock) partially melts, as viscous rock rises and pressure drops in transition from deep to shallow levels. It's mainly the pyroxene component of the peridotite that melts, forming basalt magma. The basalt rises and crystallizes to form new ocean crust in the fissure that pulls apart and marks the oceanic spreading center. As part of this melting, crystallization, and tectonism, sea water seeps downward into the system, reaching the residual olivine-rich peridotite. The olivine reacts with the sea water that has filtered into the complex and forms the hydrous minerals serpentine and chlorite (chapter 4). Now, there is a chlorite + serpentine-rich layer under the basalt ocean crust headed across the ocean and into the subduction zone. At a certain depth in the subduction zone, the serpentine and chlorite, at increased pressure, react back to olivine plus water. The water rises into the overlying mantle peridotite, eventually reaching pressure-temperature conditions where the pyroxene and olivine of the peridotite combine with water to melt into basalt magma. All told, a fascinating process linking chemical reactions and plate motions.

There is still more to the process in forming an arc volcano. We could think that the basalt magma would just shoot up to the surface, piling up into a volcano, as it does in the ocean-island volcanoes (Hawaii). But, the basalt magma in arc volcanoes typically lingers at the base of the crust and is split into different types. Pools of magma at depth in the arc begin to fractionally crystallize. Crystals of pyroxene and olivine that are richer in magnesium and iron, poorer in silica than the magma as a whole, crystallize first and settle. These crystals can form a residue on the floor of the magma chamber—a "cumulate". The remaining magma rises to the surface, poorer in Mg, Fe, and richer Si, and crystalizes as various forms of granitic rock (commonly quartz diorite) or volcanic rock (Fig. 1-6).

Arcs develop both on oceanic crust and continental crust. For our discussion of the Coast Plutonic Complex arc, we are talking about a continental arc. The interaction of arc magma and the "countryrock" that is intruded is a major issue in understanding mountain building of the Cascades.

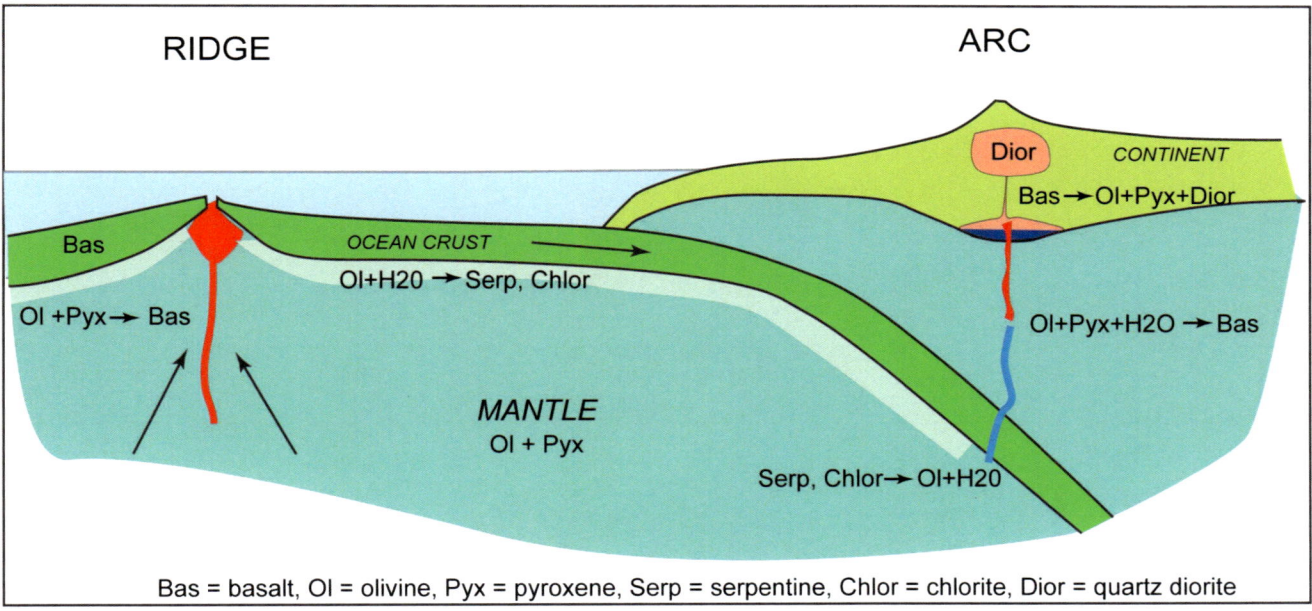

Bas = basalt, Ol = olivine, Pyx = pyroxene, Serp = serpentine, Chlor = chlorite, Dior = quartz diorite

Fig. 1-6 Plate tectonic and geochemical model linking arc magmatism to sea-floor spreading and subduction. At depth below the oceanic ridge system, mantle peridotite partially reacts with sea water, converting olivine to the hydrous minerals chlorite and serpentine.

The chlorite and serpentine are carried in the oceanic plate away from the ridge, eventually to depth in the subduction zone, where under high pressure, they react back to olivine plus water. The water streams upward, and along the active conduit causes partial melting of mantle rock to produce basalt magma that in turn rises into the crust.

Ponding at the base of the crust, the basalt magma fractionates into crystals of olivine plus pyroxene that settle, making a "cumulate", and granitic magma (commonly quartz diorite) that rises into the upper crust forming arc plutons and volcanic rock.

Rodinia

Finally, in our consideration of plate tectonics, we note that part of the story of mountain building in the Pacific Northwest begins with the "supercontinent" Rodinia (Fig. 1-7), existing from 1.3 - 0.9 billion years ago. Laurentia, the Precambrian core of modern-day North America, was surrounded by the other continents on the earth. On its west side Laurentia was snuggled in against Australia and Antarctica. This positioning, discussed in Chapter 12, explains how sediment from Australia may have ended up in the Cascades. As Rodinia broke up, the ancestral Pacific Ocean was developed and has bounded the western flank of North America ever since, albeit with much tectonic complexity related to the geologic flotsam and jetsam brought in by plate convergence.

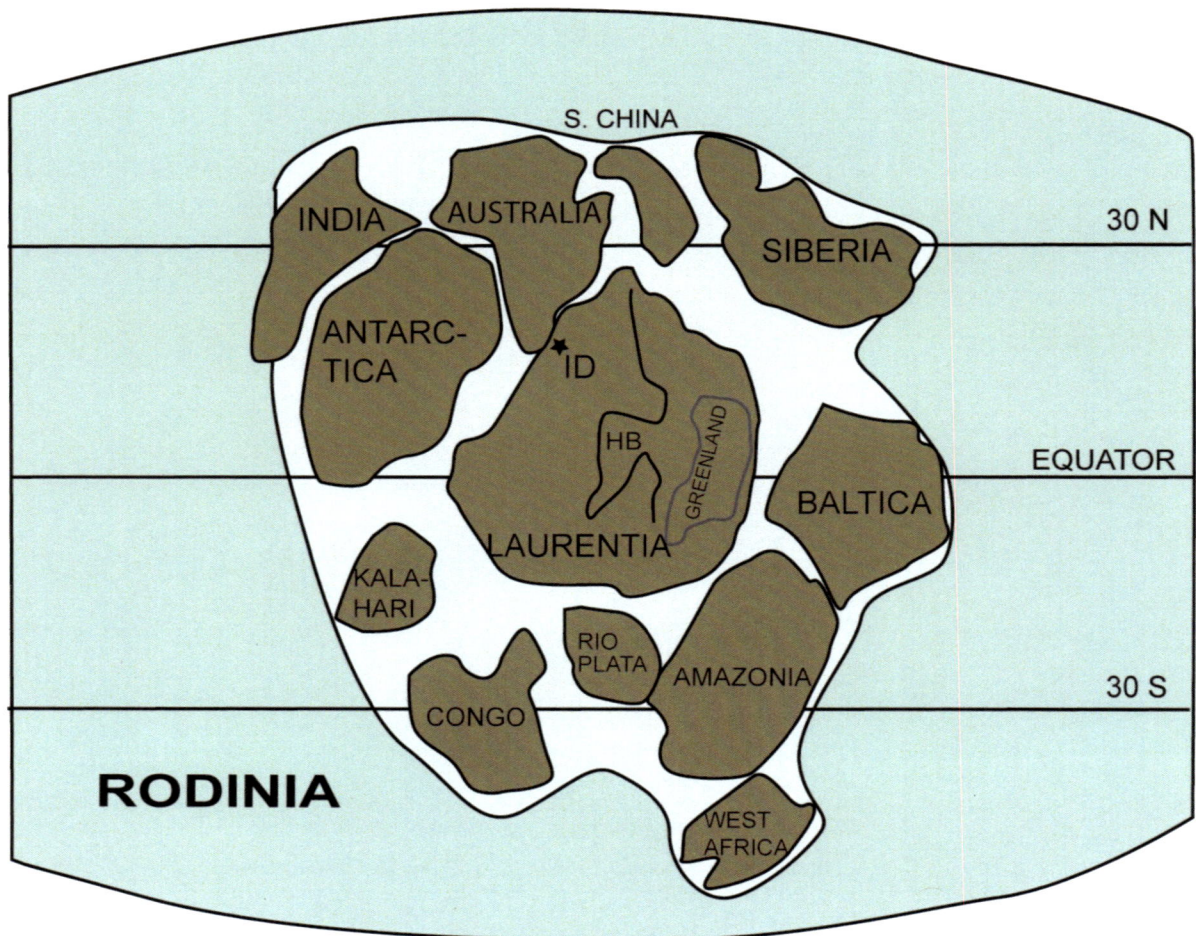

Fig. 1-7. Configuration of the supercontinent Rodinia, at about 1.3 billion years ago. ID represents the position of the Idaho region, at the edge of Laurentia, and in position to receive 1.4 billion-year-old sediment from Australia—discussed in Chapter 12. For geographic reference, HB represents the location of the modern day Hudson Bay.

CHAPTER 2
ORIGIN OF MOUNTAINS

Continents and ocean basins are fundamentally different in structure and origin. Ocean crust, as we have seen in consideration of plate tectonics, is a relatively thin sheet, three to six miles thick, of crystallized basalt magma erupted along ridge systems. Continental crust is thicker, 10 to 30 miles, and is made largely of granitic and metamorphic rock. How did this differentiation come to be? How is continental crust formed? Mountain building has created the basement to virtually all of the earth's continental landmass.

Going back to the earth's earliest days, as the planet was assembled from accreting cosmic chunks, basalt crust formed across the globe; there were no continents. Plate tectonics got going. Basalt welled up under ocean ridges, and newly generated ocean crust moved away from the ridges and sank back into the mantle in subduction zones. This was a type of convective overturn, just as a cooking pot of soup rises in the middle and sinks on the margins. The critical continent-making process happened directly above the subduction zones. Here, the over-riding plate was infused with fluids released in the subduction zone. The fluids enabled partial melting that created relatively silica-rich magma that in turn rose and crystallized as granitic rock below the earth's surface or as volcanic rock on the surface. This is the domain of the magmatic arcs described in Chapter 1. We'll come back to details of this process, which is very much fundamental to mountain origins in the Pacific Northwest, in Chapters 11 and 12.

Mountain building is in most cases a consequence of plate collision. Besides producing silica-enriched magmas in arcs (Fig. 1-6), the colliding plates commonly generate a lot of folding and faulting of rock, scrunching up the crust into a thickened welt. Some originally shallow crustal rocks become deeply buried and are metamorphosed, developing a complete make-over in mineralogy and texture.

The mountain-building process described above, involving magmatism, folding, faulting, and metamorphism in belts of plate convergence, is termed *orogeny*, from Greek roots meaning mountains and origin.

The continents as we find them now are virtually *everywhere* underlain by these belts of granitic and

Fig. 2-1. Inner Gorge of the Grand Canyon, where the continental basement of ancient metamorphic rock (schist) and granite are exposed. The orogeny creating these rocks involved arc magmatism and plate tectonic collisions ~1400-1800 million years ago (Ma). Mountains were built. Then, over a time span of nearly a billion years, the mountains were eroded almost flat, and were covered by sediment. The Tapeats sandstone capping the basement in this view is ~ 500 Ma, with the depositional contact marked by a white arrow on the left photo, and a white dashed line on the right.

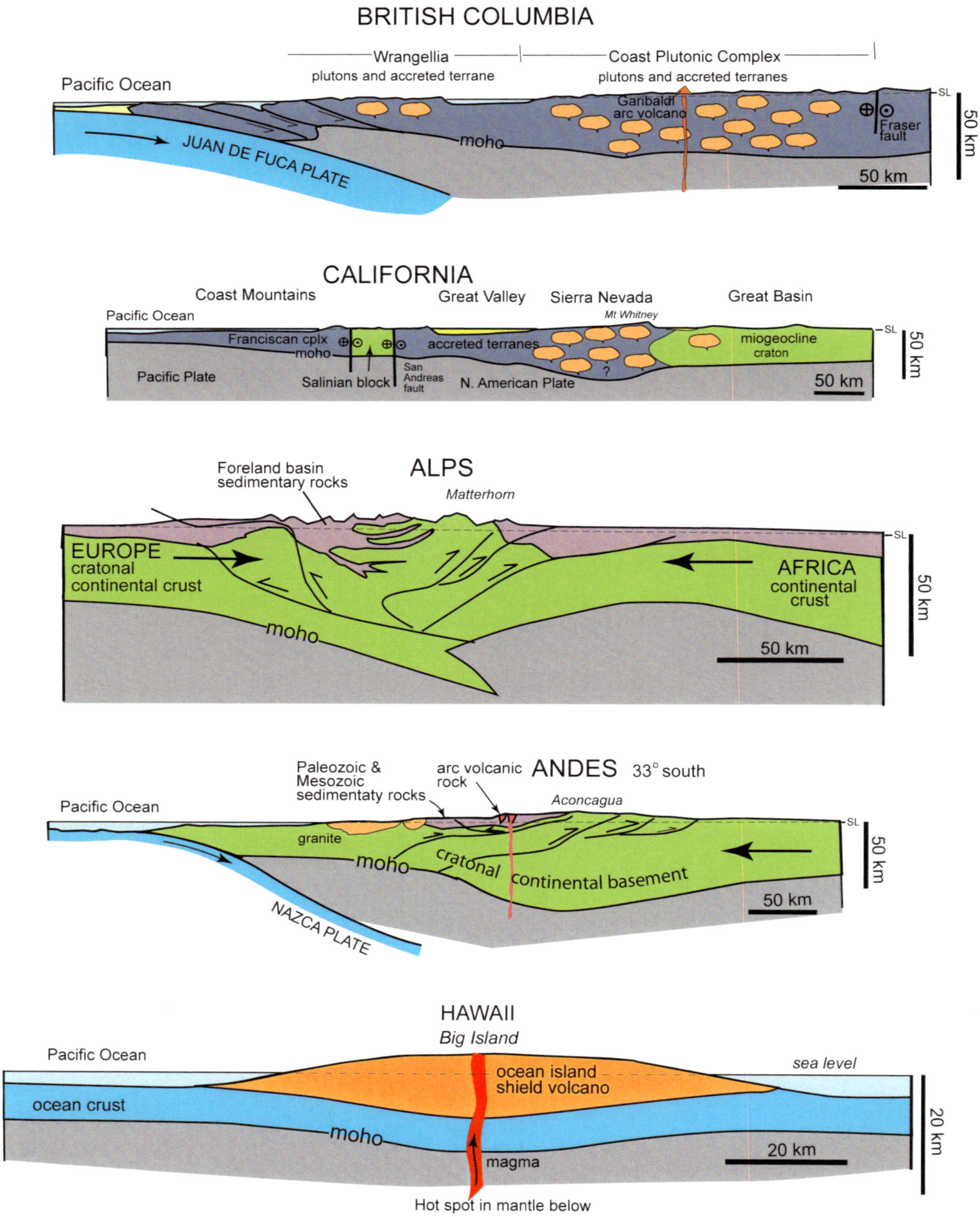

BRITISH COLUMBIA

Wrangellia — plutons and accreted terrane

Coast Plutonic Complex — plutons and accreted terranes

Pacific Ocean

Garibaldi arc volcano

moho

Fraser fault

JUAN DE FUCA PLATE

SL

50 km

50 km

CALIFORNIA

Coast Mountains   Great Valley   Sierra Nevada   Great Basin

Pacific Ocean

Mt Whitney

Franciscan cplx

moho

accreted terranes

miogeocline craton

Pacific Plate

Salinian block

San Andreas fault

N. American Plate

SL

50 km

50 km

ALPS

Foreland basin sedimentary rocks

Matterhorn

EUROPE cratonal continental crust

AFRICA continental crust

moho

SL

50 km

50 km

ANDES 33° south

Paleozoic & Mesozoic sedimentaty rocks

arc volcanic rock

Aconcagua

Pacific Ocean

granite

moho

cratonal continental basement

SL

NAZCA PLATE

50 km

50 km

HAWAII
Big Island

Pacific Ocean

sea level

ocean crust

ocean island shield volcano

moho

magma

20 km

20 km

Hot spot in mantle below

Fig. 2-2. Cross sections for selected mountain ranges on the earth. See text for discussion. Sources: British Columbia, Varsek et al. 1993; California, Mooney and Weaver 1989; Alps, Schmid 1996; Andes, Ramos et al. 2004; Hawaii, Hill 1969.

metamorphic rock. These old orogenic rocks are pretty much completely beveled by erosion. A fine display of this relationship is in the inner gorge of the Grand Canyon (Fig. 2-1). Here, because of regional uplift, the Colorado River has cut down through all the sedimentary cover of the continent in this area, exposing the orogenic basement of schist and granite that formed by plate collisions 1400-1800 million years ago. The overlying Tapeats sandstone was deposited by streams flowing across the long-since eroded mountain belt.

Because understanding of mountain building orogeny is so basic to knowledge of the earth's history, and probably because of its intrinsic complexity and challenge, and not the least the inviting setting (work and fun), a lot of geologists have devoted their careers to this enterprise.

As an introduction to thinking about the origins of mountains in the Pacific Northwest, let's take a brief

look at the structure and formation of other mountains on earth for comparison. A universal property is that a mountain pile-up of rock forms a thickened crust and causes depression of the Moho and the lithospheric mantle below. A bulge downward into the plastic asthenosphere is formed—the root of the mountains. Like a block of wood on water, a type of floating equilibrium is established. The thicker the crust the deeper the root.

Fig. 2-3. Active lava flow on the flank of Kilauea volcano, Hawaii. Ned testing.

Hawaii exhibits this process (Figs. 2-2, 2-3). As magma from the mantle "hot spot" builds up the volcano on the ocean floor, the Moho and underlying mantle lithosphere are depressed. This is an equilibrium configuration. Eventually, the volcanic island moves off the hot spot, volcanism ceases, and erosion begins to wear away the volcanic rock. With a lessened load, the crustal root of the eroded volcano rises. Now Hawaii is not the type of mountain that we are considering for comparison with the Pacific Northwest, which involves arc magmatism and plate collisions. But the formation of a crustal root and the floating equilibrium are the same process. Mountains are built up, the Moho is depressed, the building process slows or ceases, the mountains float up, and erosion strips off the upper crustal parts revealing the deep structure.

What are the main ingredients and events of orogenic

Fig. 2-4. Sierra escarpment along the east side of the range, showing 9000' of relief from valley floor to the Sierra crest. Mt Tom, el. 13,652', at left. Pine Creek moraine, deposited from a Pleistocene valley glacier, is in the center.

21

mountains? In western North America, exotic tracts of rock (terranes), mostly old offshore arcs, like the modern Japanese or Philippine islands in the western Pacific Ocean, were accreted along a lengthy portion of the continental margin, at about 170 Ma (some earlier). Thus was the initiation of the North American Cordillera and the westward building out of North America as much as hundreds of miles from the shores of the ancient continent Laurentia (Fig. 4-1). The mountain-building geology of the Pacific Northwest and California (Figs. 2-2, 2-4) reflects this process. The accreted terranes bear metamorphic and structural evidence of burial and strain inherited from collision and subduction.

In an extensive belt inboard (~200 miles) from the present continental margin, the older accreted terranes of the Pacific Northwest and California were massively invaded by arc magmas, fed by Farallon Plate subduction, in the time range of 170-50 Ma. Large bodies of intrusive igneous rock, termed plutons, were developed. Crustal thickening, deformation, and metamorphism happened and was followed shortly thereafter (millions of years) by erosion and uplift. In the Pacific Northwest, a remnant of the Farallon Plate, the Juan de Fuca Plate, is still subducting under the continent. In California most of the continental margin is the Pacific Plate, as the Farallon Plate was entirely subducted by 30 Ma.

Fig. 2-5. Torres del Paine, Chile. These glacially carved rock towers expose a horizontally layered granitic complex in the southern Andes mountains of Patagonia. Peter Gove photo.

The Andes have a plate tectonic setting much similar to the North American Cordillera. An old continental landmass, the craton, was and still is in collision with the Farallon Plate and its remnant the Nazca Plate. But, there are huge differences. The mountains, as far down in depth as one can tell, are made of the old continental crust. Terrane accretion is not part of the process. Granitic plutons are present locally (Fig. 2-5), but are mostly scarce. Arc volcanism is active, but seemingly minor in building the mountains.

So, why are the Andes there? The apparent cause is a great thickening of continental crust made by stacking of thrust sheets derived from the continent itself (Fig. 2-2). The thrusts are mapped as having moved east, away from the plate boundary, toward the interior of the continent. They formed in response to collision of the continent with the incoming oceanic plate. The collision zone itself differs from that of the North American Cordillera in that there is not now (or ever was?) a thick section of accreted oceanic rocks. In addition, the trench does not have fillings from erosion of the mountains—apparently because rain runoff is minimal in the desert climate. The down-going ocean plate scrapes directly against rocks of the craton (subduction erosion). This is an orogen greatly different from that of the North American Cordilleran margin.

Next we'll go to the European Alps (Figs. 2-2,), probably the most complicated and best known mountain geology on the earth. The setting here again is collisional, but a large oceanic plate is not part of the story. The event was basically the African Plate moving north against the Eurasian Plate. Convergence began about 120 Ma when small ocean basins of the Tethyan Sea and fragments of continent began to be subducted and accreted, both to the north under Europe and south under a small bordering plate of Africa, "Apulia". Sea floor rocks and continental crust were deeply subducted and eventually brought to the surface, some having been pushed down some 60 miles, then came up (Fig. 2-6). This process carried on and pretty much built the Alps as we see them now by 5 to 10 Ma. The plate boundary is still active in the Mediterranean, feeding the dangerous volcanoes of Etna, Vesuvius, and Stromboli.

The European Alps were made by thrusting, mostly of old crustal rocks. Arc volcanic and plutonic rocks are rare, in contrast to the great abundance of arc rocks in the Pacific Northwest and California, which attests to the long-continued subduction of the Farallon Plate.

Fig. 2-6. Hannes Hunziker of the University of Bern holds a giant white garnet crystal (Mg-rich) from the Dora Meira nappe of the western Alps in northern Italy. The host rock is metamorphosed sedimentary basement rock, originally part of the African continent. High-pressure mineral inclusions in the garnet, dated at 40 Ma, indicate burial to 60 miles in a collision zone during the Alpine orogeny. Mineral regression to lower pressure types during uplift was apparently prevented by their encasement in the strong garnet, acting as a pressure vessel. 1987.

Sedimentary deposits accumulated in small seas along the European and African margins as crust was depressed due to the increasing thickness of nappe piles. Such sedimentary rocks, related to thrusting, are termed foreland basin deposits. The sediments became rocks, and in time were caught up in the nappe formation. Such sedimentary rocks in the Alps have a notably different origin than far-travelled oceanic sedimentary rocks in the accreted terranes we see in the North American Cordillera.

What is the take-away from our view of various mountain belts? Crustal thickening is a prevailing process. Thickening happens in ocean-continent and in continent-continent collision zones. Thickening by compression, i.e. folding and thrusting, is important. But arc magmatism also contributes to thickening of crust. In the Pacific Northwest, arc magmatism is intricately involved in the mountain-building process, much more so than in some other orogens we looked at. One theme of this book is evaluation of whether crustal thickening in the mountains of the Pacific northwest was caused primarily by thrusting, or by magma emplacement.

Another theme relates to unravelling the process of subduction and accretion. In the Pacific Northwest there are broad tracts of rock that bear a history of subduction, deep burial, and uplift. We can work out a history of how the continental margin grew outward with a combination of: age-dating, mineralogic evidence of pressure-temperature conditions, and analysis of the structural imprint on the rocks inherited from their travels across ocean basins and into and out of subduction zones.

CHAPTER 3

TOOLS

Measuring the Depth of Rock Burial and Uplift

The goal of unravelling mountain building depends hugely on gaining information on how deep the rocks were buried, and how hot they became deep in the earth's crust. Minerals are sensitive to pressure, and thus provide this information. An extreme example is a diamond-bearing rock at the earth's surface. We know from experiments that the diamond, formed from carbon, was stable at pressures ~80 miles down. At shallower levels, graphite is the stable form. Rock uplift was 80 miles, and quick enough that the diamond was unable to convert to graphite. Rarely, diamonds occur in metamorphic rock derived from sediment, and in these special rocks the history includes not only 80 miles of uplift, but a preceding 80 miles of depression—-80 miles down followed by 80 miles up. That's some serious plate tectonics! On a smaller, but still dramatic scale, minerals in the Cascades and Coast Mountains provide the same kind of burial and uplift information.

Most rocks of interest in mountain building in the Pacific Northwest are igneous and metamorphic. The igneous are largely granites crystallized from magma intruded upward from depth. Metamorphic rocks are recrystallized from other rocks, most originally formed near or on the earth's surface and were transformed by heat and pressure. Minerals in the metamorphic rocks—their composition, age, and history—are a critical key to understanding how the mountains were built.

What is the metamorphic process? Consider the schist shown in Fig. 3-1. This rock, from the heart of the B.C. Coast Mountains, is part of an extensive thick layer of originally sedimentary mud and sand deposited on the sea floor marginal to the edge of North America about 130 million years ago. We think

0.5 inch

Fig. 3-1. This schist, in the B.C. Coast Mountains, was created from mud on the ocean floor. Deep burial in the earth caused heat and pressure that totally rebuilt the rock into the schist we see now. The dark brown mineral is staurolite, the red equant mineral is garnet, and the whitish prismatic mineral is kyanite. The groundmass in this rock is made of quartz, plagioclase, and biotite.

we can track this metamorphic rock into unmetamorphosed sedimentary equivalents—-sandstone, and shale. The schist is completely recrystallized into a different set of minerals and is sporting a totally new fabric (mineral alignment). To understand this process of metamorphism we need to visualize the original sediment as a pot of chemical elements: Si, Al, Fe, Mg, Mn, Na, K, Ca, O, H. These elements occur as ions: 1) bound up in crystals, e.g. quartz is SiO_2, or 2) dissolved in water in the sediment, as in Si^{+4}, O^{-2}. Metamorphism pretty much totally reorganizes the combinations of ions in a parent rock into different minerals in the metamorphic rock. What force of nature makes this metamorphism happen? The answer comes down to the science of thermodynamics, and explains why geology students interested in metamorphic rocks end up taking advanced courses in physical chemistry.

The basic theory is not that hard to understand. The soup of elements in a rock responds to changes in pressure and temperature. When pressure goes up, the elements will reorganize from one set of minerals to another that is more compact. Diamond and graphite are a good example: diamond density is 3.5 gm/cm^3, graphite is 2.2 gm/cm^3. Diamond is the high-pressure form. Nature makes diamond from graphite by deep burial; humans can replicate this process by squeezing graphite in huge presses.

The relationship of temperature to mineral change is less obvious. Here we are talking about degree of disorder ("entropy" in thermodynamic lingo). A more ordered state for ions corresponds to a lower temperature state. The melting of solids with increased temperature exemplifies this relationship. In solid form, a mineral has a tight, ordered internal lattice of ions bound together. In the liquid form, these ions are on the loose, flowing and sliding past one another— much less order. The change from liquid to vapor is similar in terms of increased entropy following rise in temperature. Not so obvious is the change of one mineral to another caused by temperature change; entropy is a function of order within the crystal lattice.

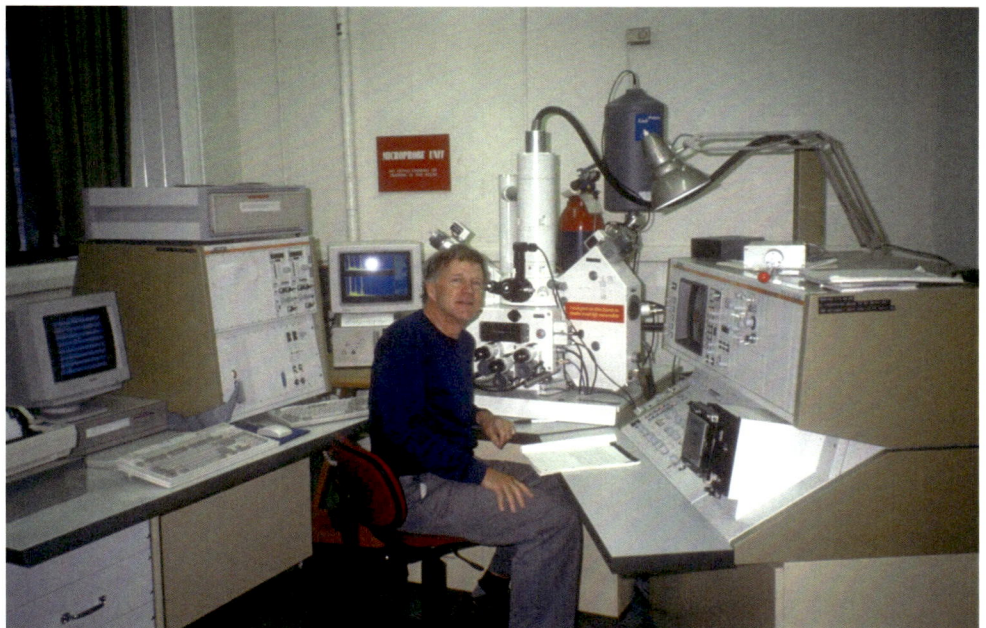

Fig. 3-2. Ned analyzing the chemical composition of minerals with the *electron microprobe* at the University of Otago in New Zealand, 1995. A thin slice of rock about an inch across is mounted on a glass slide, polished, then placed in the vacuum chamber at the base of the microprobe column. Then, an electron beam accelerated down the tube zaps the mineral sample in a spot a few microns wide. X-rays specific to the individual elements in the mineral are generated, spread out, and are collected and counted by detectors. The x-ray counts are then quickly worked over by a computer that spits out the element composition of the mineral. Quite a miracle!

The beauty of these thermodynamic properties is that given a soup of elements, the way they organize is a function of temperature and pressure.

Experimental and theoretical studies have quantified these relationships. All we need to do to get the T and P is identify the minerals and measure their compositions, as with the electron microprobe (Fig. 3-2).

Mineral formulas

Quartz SiO_2

Chlorite $(Mg,FeAl)_6 (SiAl)_4O_{10}(OH)_8$

Biotite $K_2 (Fe^2, Mg)_3 AlSi_3O_{10} (OH)_2$

Muscovite $K_2Al_2 AlSi_3O_{10} (OH)_2$

Garnet $(Fe^2, Mg, Ca, Mn)_3 Al_2 (SiO_4)$

Kyanite, Sillimanite, Andalusite Al_2SiO_5

Staurolite $Fe^{2+}_2Al_9O_6 (SiO_4)_4(O, OH)_2$

Plagioclase

$NaAlSi_3O_8$ albite end member

$CaAl_2Si_2O_8$ anorthite end member

Crossite $Na_2(Mg,Fe^2)_3(Al,Fe^3)_2Si_8O_{22}(OH)_2$

Actinolite $Ca_2(Mg,Fe^2)_5Si_8O_{22}(OH)_2$

Hornblende $(Ca,Na)_{2-3}(Mg,Fe^2,Al)_5(Al,Si)_8O_{22}(OH)_2$

Epidote $Ca_2(Al,Fe^3)_3(SiO_4)_3(OH)$

Olivine $(Mg,Fe^2)SiO_4$

Pyroxene $(Mg,Fe^2)SiO_3$ enstatite

Serpentine $Mg_3(OH)_4Si_3O_5$

Chromite $Fe^2Cr_2O_4$

Magnetite $Fe^2Fe^3_2O_4$

Fig. 3-3. *above.* A reference list of the main minerals of interest in this book.

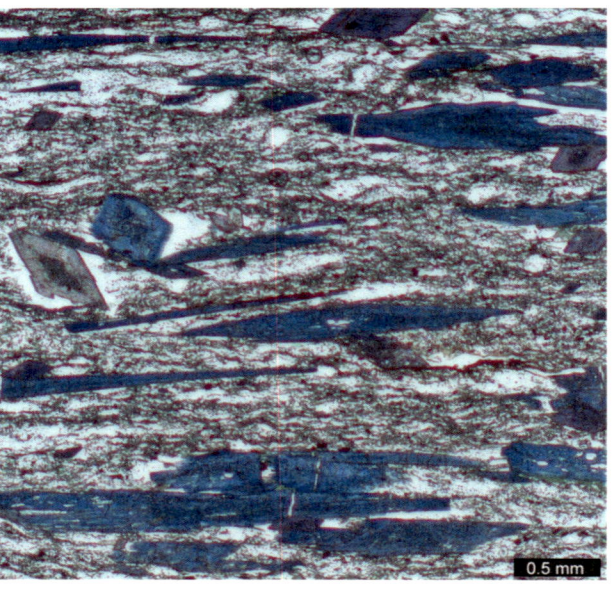

Fig. 3-5. *above.* Crossite in Shuksan blueschist, viewed through the microscope.

Fig. 3-6. *below.* Quartz (Q), plagioclase (P) and hornblende (H) in granitic rock of the Mt. Stuart batholith. Crossed polarizers in the microscope view.

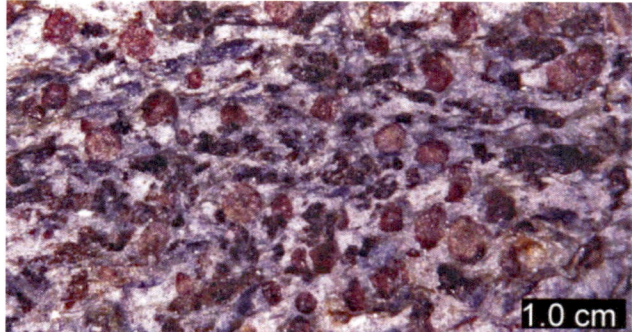

Fig. 3-4. *above.* Garnet - kyanite schist in country rock of the Breakenridge pluton, B.C. Coast Mountains. This rock has undergone a major do-over from its sedimentary parent of mud. Minerals are: garnet - red, kyanite - blue, mica and quartz - white.

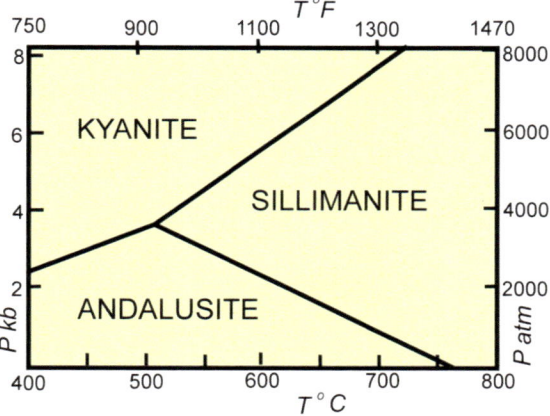

Fig. 3-7. Pressure-temperature stability of
the Al₂SiO₅ minerals: kyanite, sillimanite and
andalusite.

Common minerals referenced in this book are listed in
Fig. 3-3, and examples in Figs. 3-4, 3-5, 3-6.

How do the minerals in a rock give us the temperature
and pressure of their formation?

The three minerals kyanite, sillimanite, and andalusite
are "polymorphs"—they have the same chemical
composition but different structure of the crystal
lattice; they form during metamorphism at different
conditions of temperature and pressure (Fig. 3-7).
Collectively, we refer to these three minerals as
aluminum silicates. They are quite distinctly different
looking from one another.

In the Cascade Core, and in the B.C. Coast Mountains,
andalusite occurs rarely as unaltered "fresh" grains
(Fig. 3-8), indicating low pressure. But more
commonly the andalusite grains are replaced by
kyanite or sillimanite (Figs. 3-9, 3-10), preserving the
original andalusite shape. Also, some grains show the
distinctive cross pattern of graphite inclusions (Fig. 3-
9), which helps us know that the grain was originally
andalusite.

These replaced andalusite grains tell a story of increase
of pressure (burial), that then can be related to tectonic
and plutonic forces of mountain building.

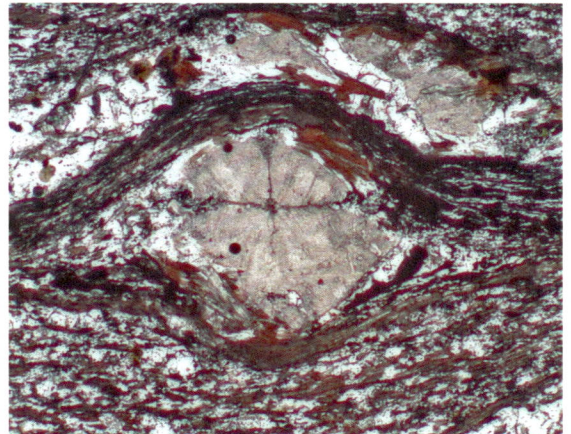

Fig. 3-8. Fresh andalusite, not altered to
kyanite or sillimanite. Note distinctive
cross-shaped dark inclusion pattern, made
by graphite grains. Microscope view.

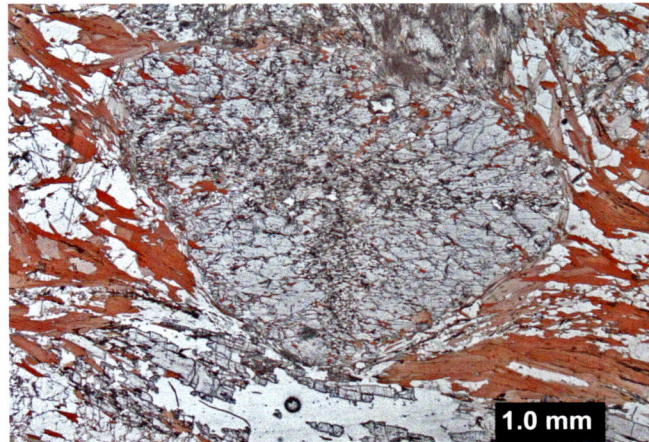

Fig. 3-9. Andalusite replaced by
sillimanite. The graphite inclusion pattern is
preserved. The view is looking down the long
prisms of sillimanite, aggregated as a mass of
diamond-shaped grains in this cross-section.

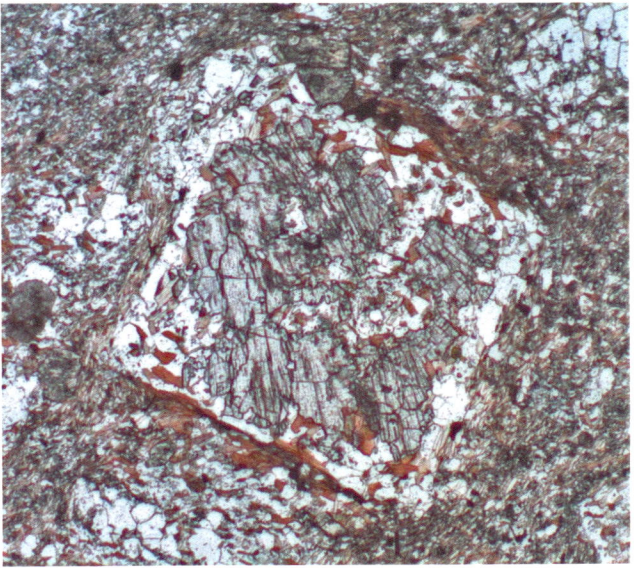

Fig. 3-10 Andalusite replaced by kyanite,
which exhibits a bladed form—different from
the diamond shapes of sillimanite in Fig. 3-9.

Combinations of mineral assemblages are of great value in determining the pressures and temperatures of rock formation. As unlikely as it seems, minerals in a rock are tuned in to each other in the sense that the ions that are combined in the lattice structure of each mineral will jump from one mineral to another under the influence of changes in temperature and pressure. As a bit more explanation, consider the formulas of garnet (Fe^2, Mg, Ca, Mn)$_3$ Al$_2$ (SiO$_{4)}$ and biotite K$_2$ (Fe^2, Mg)$_3$ AlSi$_3$O$_{10}$ (OH$_2$. In garnet the Fe, Mg, Ca, and Mn all occupy the same part in the crystal lattice. The same for Fe and Mg in biotite. We term this mixing on a lattice site "solid solution".

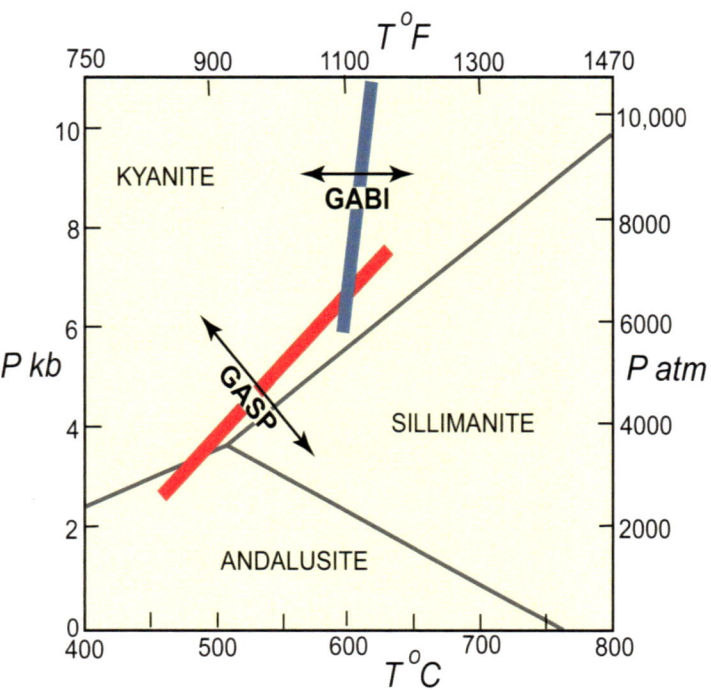

Fig. 3-11. Pressure—temperature plot showing relative stability of kyanite, sillimanite and andalusite. Also shown is the shifting equilibrium of assemblages of: garnet + biotite—GABI, and garnet+aluminum silicate+plagioclase—GASP.

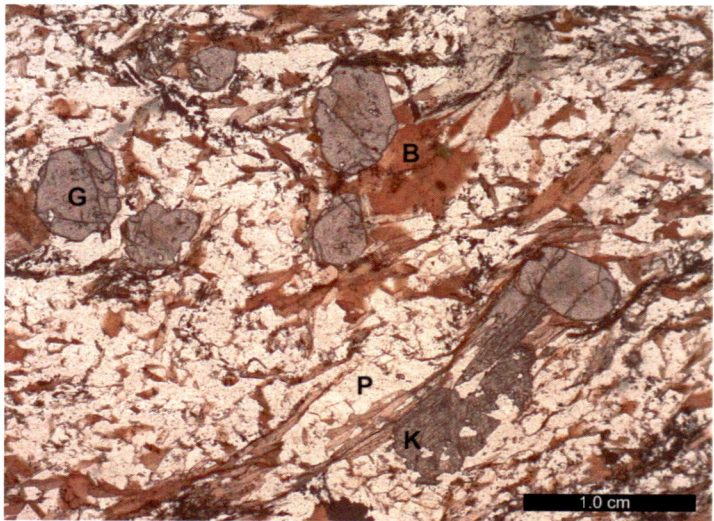

Fig. 3-12. Chiwaukum Schist, North Cascades. Garnet occurs in equant high-relief grains. Biotite is a brown platy mineral, a variety of mica. Kyanite is a mineral also of high relief, but with strong cleavage traces (cracks), and occurring as elongate and equant grains. Plagioclase is colorless.

GABI on the diagram of Fig. 3-11 refers to an exchange of iron and magnesium ions in garnet and biotite that is particularly sensitive to temperature (but not pressure). Both minerals contain magnesium and iron ions that mix in the solid solution. As the temperature rises, some magnesium in biotite moves over into the neighboring garnet, and iron moves from garnet to biotite. Thus the blue line on Fig. 3-11, that represents T and P for a certain distribution of Fe and Mg, shifts. This exchange is calibrated to T and P from lab experiments. We measure the iron and magnesium composition of the biotite and garnet with the microprobe (Fig. 3-2), and there's the position of the blue line.

GASP (garnet, aluminum silicate, plagioclase) similarly represents a shuffling of ions in response to T and P (Fig. 3-11). In this scenario, with increased pressure, or decreased temperature, plagioclase feldspar gives up some of its calcium component to increase the calcium in associated garnets. To balance this exchange, also produced are aluminum silicate and quartz. The reaction is dependent on both temperature and pressure-—the red line is sloped. With the microprobe, we measure the compositions of garnet and plagioclase, and we have the red line. (The motivated reader can balance this reaction using the mineral formulas given in Fig. 3-3).

Commonly the GABI and GASP minerals

occur in the same rock (Fig. 3-12). So, where the equilibrium lines for these two systems cross, we have the very pleasing situation of obtaining a specific temperature and pressure for the rock (Fig. 3-11).

Here's another cool feature of mineral compositions that helps us build a story of tectonics: garnets are commonly zoned. The zoning records an early stage of growth (in the core) that happened under different P-T conditions than for the rim. In the Cascades and B.C. Coast Mountains, the zoning is invariably low in Ca in the core, high in Ca on the rim. Where the garnets coexist with biotite, plagioclase, and aluminum silicate (andalusite, sillimanite, and kyanite) the zoning gives us a track of the P-T conditions during garnet growth.

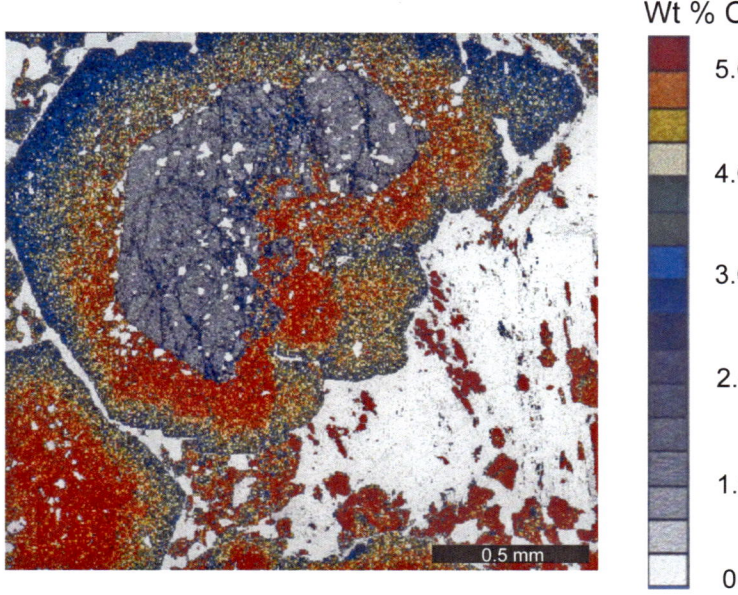

Wt % Ca

Fig. 3-13 shows a garnet mapped on the microprobe for calcium—the indicator element for pressure. These values, together with compositions for biotite and plagioclase, allow solution for P and T at the different zones within the garnet, shown on Fig. 3-14. The rim calculation, based on rims of biotite, as well as plagioclase, is the most accurate, as uncertainty arises in correlating compositions measured inside the mineral

Fig. 3-13. Zoned garnet mapped (on the microprobe) as colors representing Ca concentration. This sample is from the Nooksack-Harrison terrane in the B.C. Coast Mountains at the contact zone of the Breakenridge pluton (Fig. 11-25). Sample 179MD17.

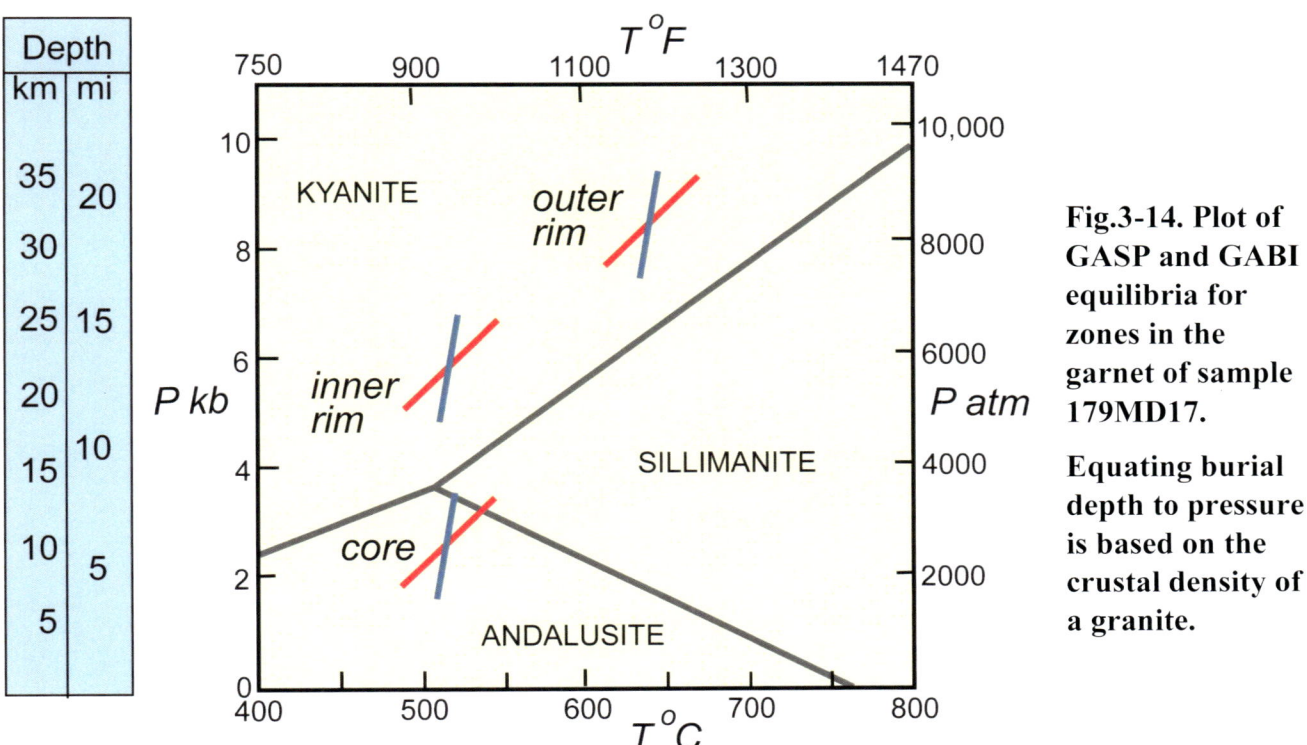

Fig.3-14. Plot of GASP and GABI equilibria for zones in the garnet of sample 179MD17.

Equating burial depth to pressure is based on the crustal density of a granite.

29

grains for the older events. The core zone formed at a depth of about five miles; the rim at 20 miles deep.

Throughout the Cascades Core and British Columbia Coast Mountains, we find in metamorphosed sedimentary rocks garnets strongly zoned in Ca. Burial as much as 20 to 25 miles during garnet growth is indicated. And of course the rocks have been uplifted the same distance. This finding of huge vertical crustal movements within the magmatic arc of the B.C. Coast Mountains and Cascades Core is stunning, and the cause of which is much debated. We revisit this finding in later chapters.

From the mineralogy, and P-T conditions of formation in the Cascades Core and in counterparts in the B.C. Coast Mountains, we find mostly higher temperature and lower pressure metamorphic conditions than in the northwest Cascades thrust system (Fig 3-15). Metamorphic rocks of the Cascade Core and B.C. Coast Mountains are closely associated with plutons. It is the "country rock" into which the plutons were intruded. The metamorphic and plutonic rocks together constitute a "continental arc" (Fig 1-2).

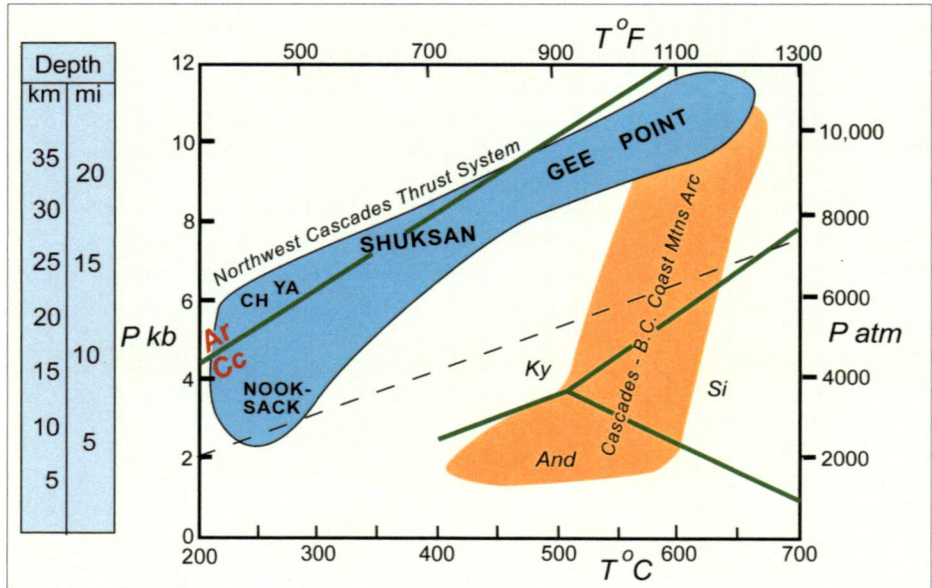

Fig. 3-15. Pressure-temperature conditions and depth of formation of metamorphic rocks in the: 1. northwest Cascades thrust system and 2. magmatic arc system of the Cascades Core and B.C. Coast Mountains. The dashed line shows a normal P-T gradient in the earth. The green line separates the stability fields of aragonite and calcite. And = andalusite, Ar = aragonite, Cc = calcite, CH = Chilliwack Group, Ky = Kyanite, Si = sillimanite, YA = Yellow Aster Complex. Gee Point refers to a high-grade zone in the Easton Metamorphic Suite, and Shuksan refers to lower-grade Easton Suite.

So, now let's look at the mineralogy of the northwest Cascades thrust system. Key minerals here indicate metamorphism at high pressure -low temperature conditions (HP-LT), the reverse of the Cascades Core. The tectonic environment envisaged is that of a subduction zone—cold crust slips down to great depth, where pressure rises quickly but temperature lags. Key minerals are aragonite, which is the high-pressure polymorph of calcite ($CaCO_3$), and crossite (blue amphibole, Figs. 3-3, 3-5).

Some rocks of the thrust system show almost no evidence of high-pressure metamorphism until they are x-rayed. But, as Joe Vance of the University of Washington discovered in the 1960s, the calcium carbonate mineral is not the usual calcite that makes up limestones, but is the high pressure polymorph aragonite. This finding is momentous in the geologic world, as the aragonite polymorph of $CaCO_3$ is stable in rocks formed only at very great depths (Fig. 3-15), and so solidifies the subduction zone interpretation.

All the nappes in the North Cascades thrust complex except the Mélange Belt have aragonite, and therefore were subducted. But, the rock unit over which the nappes were thrust—Nooksack Formation—has calcite instead of aragonite; thus we know it formed at lower pressure and was not involved in subduction. Likewise, the actual fault zones along which the nappes slid into place contain calcite;

aragonite does not occur. We conclude that the subduction event occurred prior to emplacement of the nappes.

Measuring the Age of Rocks

As important as any aspect of a rock for understanding its part in the mountain-building process is its age. A modern-day scientist with a new-fangled lab can come into a long-studied area and turn the interpretations upside down with previously unavailable isotopic ages. These tools are all based on the property of radioactive decay. A radioactive element breaks down to a decay product element, plus radiation, at a constant rate. We measure the amounts of the original element and the decay product element. Then knowing the decay rate, we can calculate the age of the sample—that is, when the decay product first formed. There are many complexities, including that for a given element of interest there are variations in atomic mass; an atom of a certain element can exist as different isotopes, same number of protons and electrons, but different numbers of neutrons. The isotopes have a distinct mass, and thus their amount can be measured in a mass spectrometer. Nowadays, many geologists working on mountain-building geology find themselves separating minerals (Fig. 3-16) for measurements of isotopic ratios in the mass spectrometer lab. Results of these measurements (Fig. 3-17) are enlightening.

Element systems that work for isotopic dating of rocks in the Cascades and B.C. Coast Mountains, given in the order of parent → decay product, and the minerals used for age determination, are:

Uranium → lead   U/Pb   zircon and monazite
Potassium → argon   K/Ar   (Ar/Ar is a variant process)   mica and hornblende
Samarium → neodymium   Sm/Nd   garnet

What do these radiogenic systems date as far as geology?

U/Pb zircon gives for igneous rocks the age of crystallization of the magma, and for sandstones the maximum depositional age of zircon sand grains, i.e. a maximum possible age of the sedimentary rock. U/Pb monazite is good for metamorphic ages.

In the K/Ar system the decay product Ar gas leaks out of the crystal at elevated temperatures in the

Fig. 3-16. Separating zircons. Step 1. Ned is crushing 20 lbs. of sandstone, Step 2. Linda Brown separates heavy from light minerals on the water table. Steps 3&4 (not shown) include the magnetic separator and heavy liquids. The final treasured concentrate of detrital zircons, far right, is about the volume of a pencil eraser.

range of : 530°C for hornblende, 350°C for muscovite mica, 280°C for biotite mica. These ages, marking the point at which the Ar is locked in, are very useful in dating uplift of deeply buried rocks. The K/Ar system is also good for dating the crystallization age of low-grade metamorphic rocks.

The Sm/Nd system, used to date garnets, gives us a direct age of peak metamorphic conditions.

Ages of rocks discussed here are millions of years old. An abbreviation of common use is Ma, which means: millions of years old. Thus, the Eldorado pluton has been dated by measuring U/Pb in zircon as 88 Ma—88 million years old.

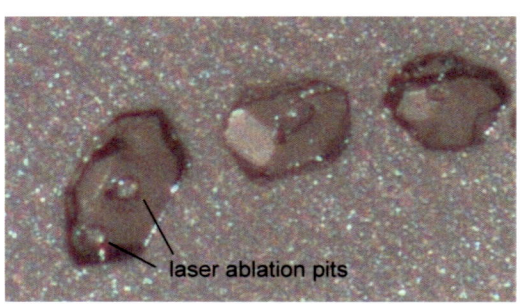

Fig. 3-17 A. Zircon grains with tiny ablation pits formed during age analysis by a focused laser beam blasting a hole and vaporizing a bit of the zircon. The zircon vapor that is created mixes with hot plasma and feeds through the mass spectrometer where isotopes are measured. The pits are about .03 milli-meters across. A hundred or more spots are commonly analyzed for each sample.

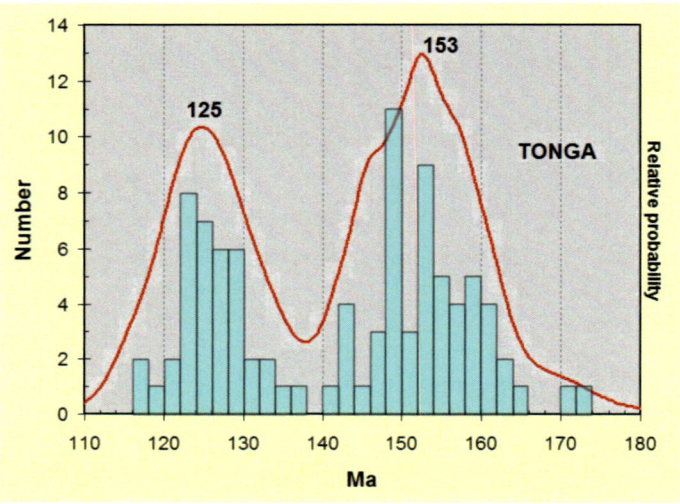

Fig. 3-17 B. Zircon ages for a single sample. Each vertical column is a measure of the number of spot ages in the time bracket on the X axis. This sample is a sand-stone, from the Tonga formation in the North Cascades, with two dominant age populations of zircon grains.

Structural symbols

On maps geologists have symbols indicating various structural features, such as faults, and the orientation of sedimentary beds or metamorphic foliations. A few of these symbols, used in this book, are shown in Fig. 3-18.

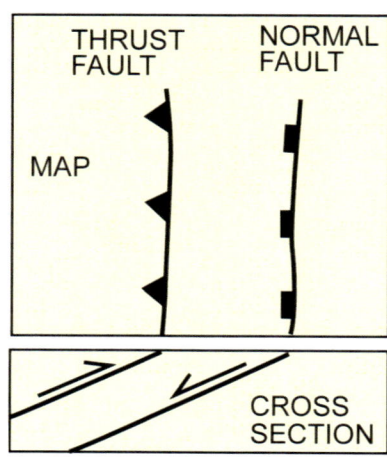

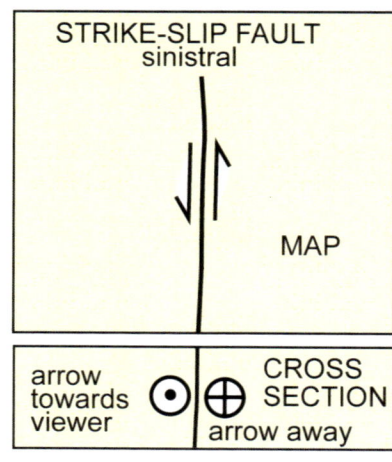

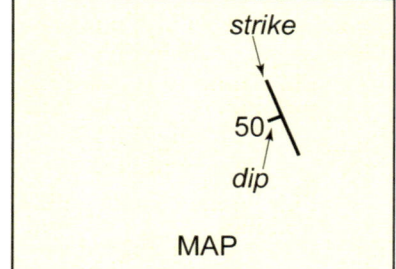

Strike gives the direction of a horizontal line lying in a geologic plane, such as bedding. *Dip* gives the angle between the geologic plane and horizontal.

Fig. 3-18. Thrust and normal faults are marked by a heavy line with symbols on the upper side. For a strike-slip fault, arrows on the map indicate the relative movement directions, which could be sinistral or dextral. Strike and dip symbols delineate the orientation of a plane.

CHAPTER 4

REGIONAL GEOLOGY

North American Cordillera

Rocks we see along the western coast of North America are of course old—tens to hundreds of million years—but they are actually young compared to the age of the continent, which goes back in places billions of years. The western part is characterized not only by relatively young rocks, but also by the formation of these rocks into mountain ranges, and is termed the North American Cordillera (Fig. 4-1).

Prior to development of the Cordillera, the edge of the continent lay far east of the present margin. In its early days ~600 million years ago, the continent had just recently split away from the supercontinent Rodinia. (Fig.1-7) An ocean opened up between North America and what is now Asia, Australia, and Antarctica. The ocean, named Panthalassa, grew for hundreds of millions of years, by sea-floor spreading as North and South America moved away from the other continents. The western margin of North America was "passive" in the sense that it was at the trailing edge of a moving plate, as opposed to "active" when the edge of the plate is collisional. The margin remained passive until about 360 million years ago, and during that time of no mountain building and relatively low topographic

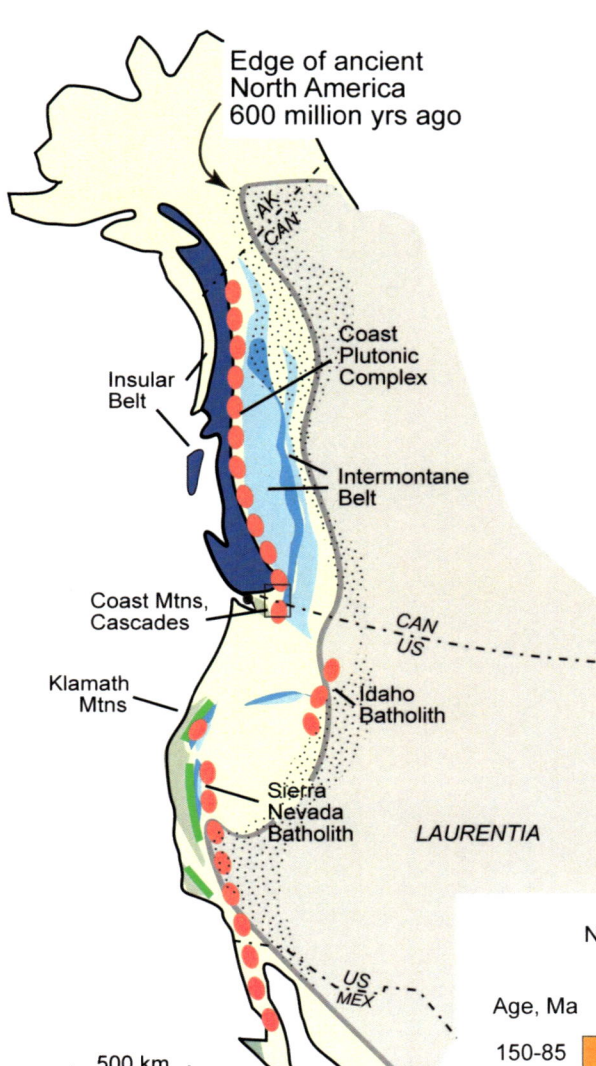

Fig. 4-1 *above*. Simplified sketch of the North American Cordillera.

***Right*. Chart of predominant terranes in the Cordillera and correlations with rock units of the Pacific Northwest.**

Age, Ma	North American Cordillera		San Juan Islands Northwest Cascades Thrust System	Southern Coast Plutonic Complex
150-85	(orange)	Great Valley, Methow fore-arc basin		***Methow-Settler-Chiwaukum***
150-75	(grey-green)	Franciscan assemblage, subduction complex	***Easton terrane***	
170-160	(green)	Coast Range ophiolite, marginal arc	***Fidalgo Ophiolite***	
235-110	(yellow)	Nooksack-Harrison, arc and back-arc basin	***Nooksack Fm***	***Harrison L. sequence***
235-170	(blue)	Cache Creek assembl. exotic ocean floor	***Bell Pass Melange***	***Bridge River***
370-170	(light blue)	McCloud assemblage, fringing oceanic arc	***Chilliwack Gp.***	***Cadwallader Chelan Mtns.***
550-350	(dark blue)	Wrangellia and Alexander terranes, exotic ocean arc		***Wrangellia***
1400-350	(dotted)	Miogeocline, passive margin sediments		
3500-600	(grey)	Laurentian craton, ancient North American continent		
170-50	●●●	Continental arc		***Coast Plutonic Complex***

relief, quiet waters deposited well-washed sands and mud along the western continental margin. These sediments comprise the "miogeocline" (Fig. 4-1), and are distinctive from the younger coarse and poorly sorted deposits in the mountain-building days that followed.

Terranes

Next in the evolution of western North America, large tracts of earth crust travelled by plate tectonics to be added to the continental margin. These mobile bodies began arriving about 300-350 million years ago, and mostly came in at about 170 Ma, growing the continent westward by accretion of many separate pieces. In working out the structures of continental growth, geologists must consider that any given body of rock could be a separate entity, having travelled and accreted by itself. The term terrane is used to acknowledge possible separateness, based on distinctive geology and faulted boundaries. Many distinct terranes make up the Cordillera (Figs 4-1, 4-2).

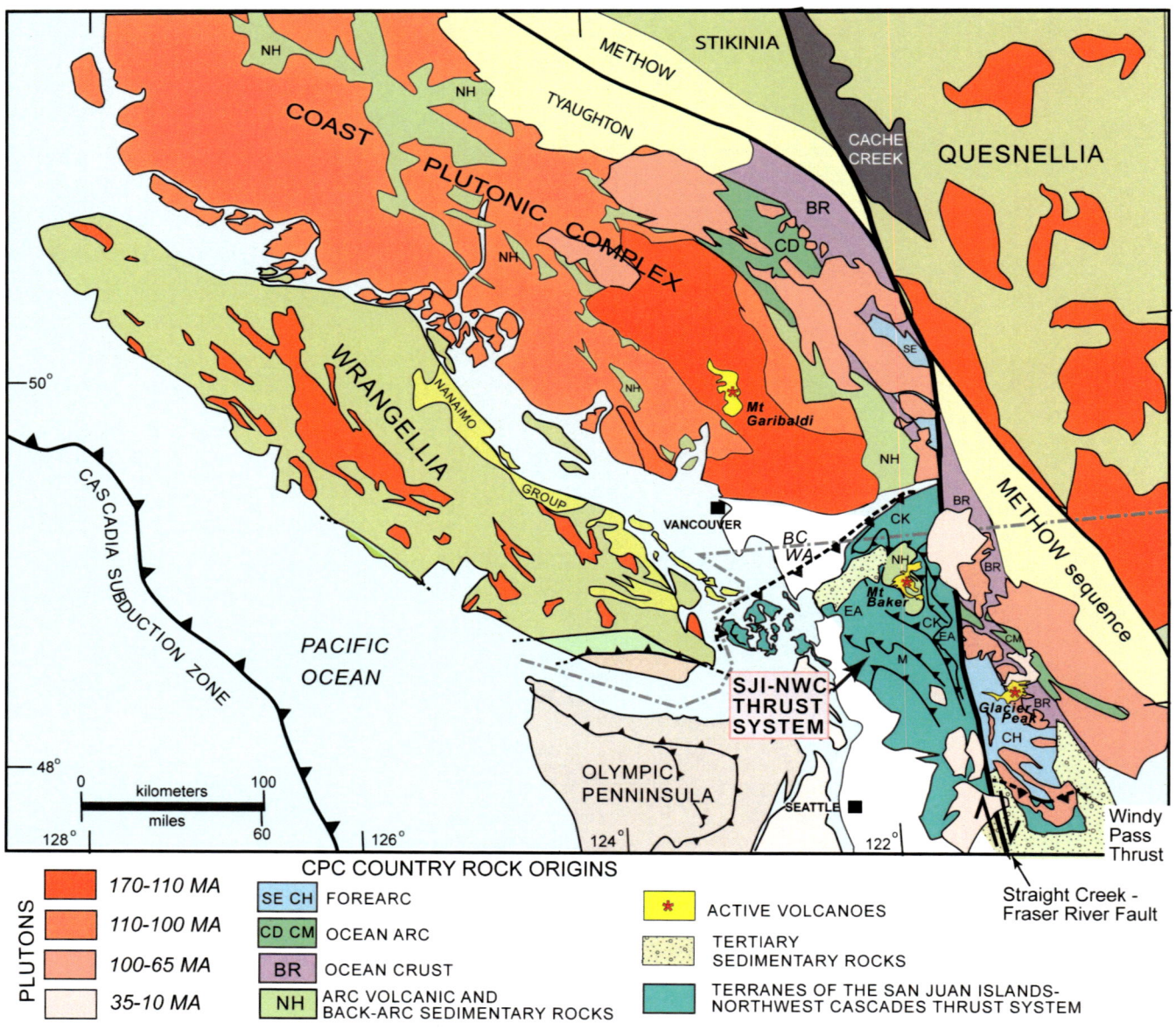

Fig. 4-2. Map of Pacific Northwest bedrock geology with emphasis on terranes involved in orogeny of the Coast Plutonic Complex (CPC) and the San Juan Islands-northwest Cascades thrust system. Terrane abbreviations: BR=Bridge River, CD=Cadwallader, CH=Chiwaukum, CK= Chilliwack, CM=Chelan Mtns., EA=Easton, M=Melange, NH=Nooksack-Harrison, SE=Settler.

34

The history and structural means by which the terranes accumulated along the Cordillera present an amazing story. All this happened very slowly by plate tectonics, but over tens to hundreds of millions of years, great displacements occurred.

The accreted terranes are virtually all of oceanic origin, which makes sense in that the incoming plate was from the ocean, as we saw in Chapter 1. The rocks formed in a variety of ocean environments: ocean crust developed at spreading ridges, ocean island volcanoes over "hot spots" as in the modern-day example of Hawaii, oceanic arcs marginal to a continent, and oceanic sedimentary basins near continental or oceanic arcs (Fig. 1-2). Rarely, we find an accreted terrane with a true continental origin; an intriguing example is the Yellow Aster Complex in the North Cascades (Chapter 9).

The terranes have distinctive histories of origin and transit through the oceanic realm. Even after accretion to the continent, terranes were shuffled.

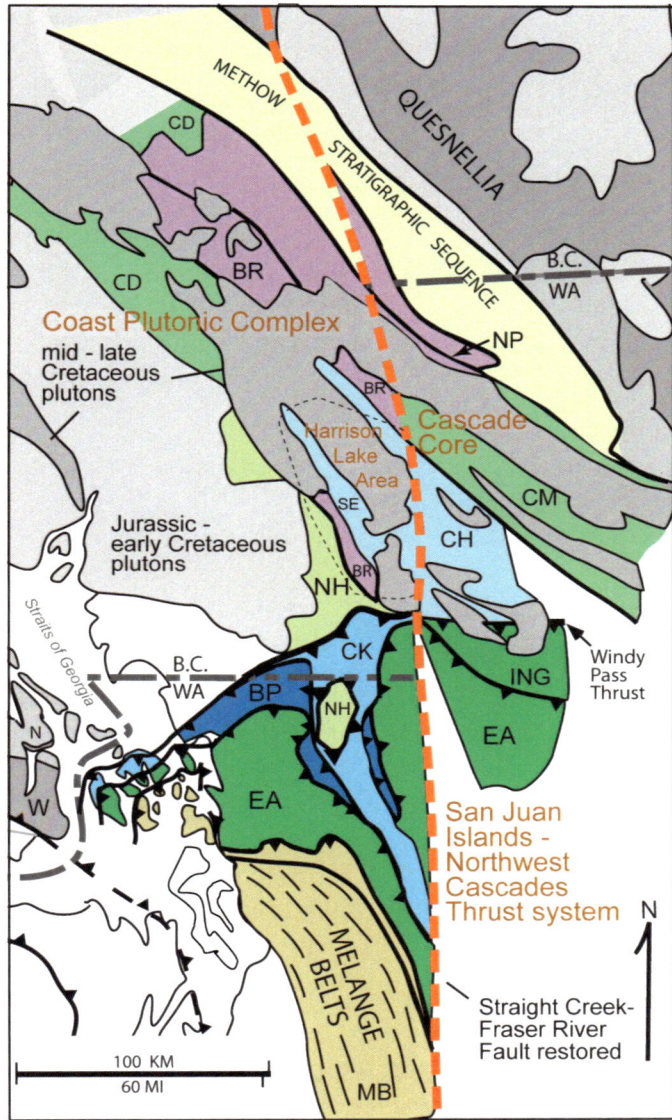

Fig. 4-3. *Left:* Pacific Northwest geology prior to displacement on the Straight Creek -Fraser River Fault at ~ 47 Ma.
Terrane abbreviations: on Fig. 4-4.

Fig. 4-4. *Below.* Summary outlining the countryrock terranes and their mutual structural relationships, prior to the Straight Ck - Fraser River Fault. The structures break into three major packages faulted together:
1. Terranes marginal to the continent by 170 Ma, accreted by obduction, and not much displaced since that time. These units are country rock to the CPC plutons.
2. Sedimentary and volcanic rocks formed along the inboard margin of the early, northern, part of the Coast Plutonic Complex; the rocks were translated south some hundreds of miles.
3. The San Juan Islands-northwest Cascades thrust system; after subduction, terranes travelled north and were emplaced over the south end of the Coast Plutonic Complex.

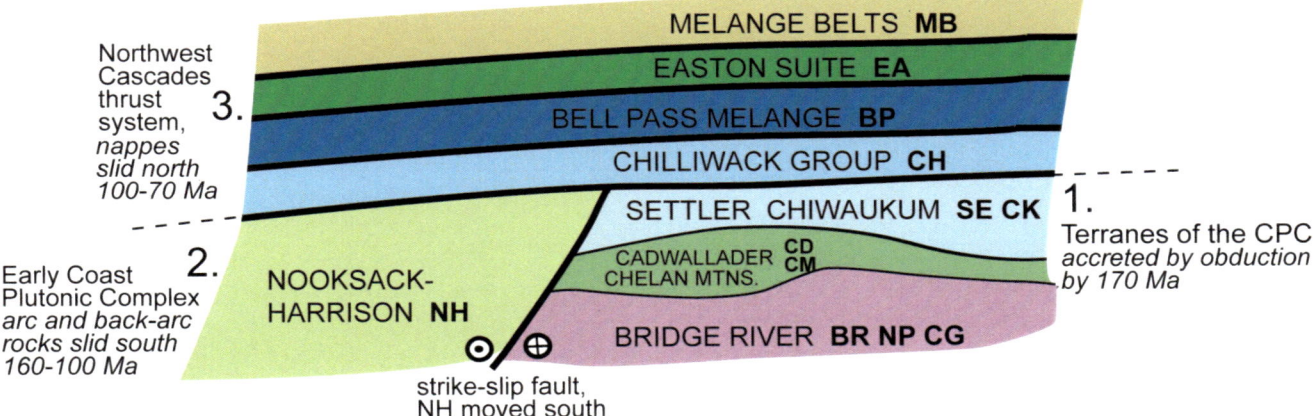

What we see now are three groups that came to their final resting place by different tectonic events, best visualized by restoring the relatively young Straight Creek - Fraser River fault (Figs. 4-2, 4-3, 4-4).

The first event was accretion of the Quesnellia and Stikinia terranes, ocean arc formations that were emplaced over the western edge of the ancient North America (Laurentia). Also at about this time (160 - 170 Ma), Wrangellia came aboard of the westward growing continent in the area of northern British Columbia and southern Alaska

Next was the initiation of the continental arc of the Coast Plutonic Complex—massive intrusions of granitic plutons along the coastal margin and inland, hosted by country rock of Wrangellia and Stikinia/ Quesnellia. The axis of the arc migrated east with time.

The Coast Plutonic Complex arc was sliced along a north-south strike-slip fault, moving the older plutons and related volcanic and sedimentary rocks along the coast south into our neighborhood.

Finally, the last big terrane shuffling event was arrival from the south of the stack of nappes forming the San Juan Islands - northwest Cascades thrust system. These oceanic rocks have a history of accretion by subduction somewhere south along the continental margin. They were exhumed, translated north and thrust over the south end of the Coast Plutonic Complex.

Terranes of the Coast Plutonic Complex

Let's first look at the major regional terranes in the Coast Plutonic Complex (CPC), which are Stikinia, Quesnellia and Wrangellia (Fig. 4-2, 4-3, 4-4). These accreted by about 160-170 Ma, emplaced as far as we can tell by obduction—pushed up over the edge of North America. Stikinia and Quesnellia are ancient island arc systems, 370-180 Ma old. A distinctive fossil assemblage, known as the McCloud fauna, bears some commonality with continental species; thus, these terranes are thought to have formed in an oceanic setting not far from the North American margin—a "fringing arc".

Wrangellia (Figs. 4-1, 4-2) is a component of the Insular superterrane, an elongate outermost terrane of the northern Cordillera, extending from Vancouver Island to Alaska. It is composed mostly of ocean arc rocks, but unlike Quesnellia and Stikinia it does not bear evidence of proximity to western North America during a long history at sea, 570-190 Ma. This terrane accreted along the northern coast (near Prince Rupert and farther north) by about 160 Ma, and took up the role of "country rock" to plutons of the Coast Plutonic Complex. In turn, the CPC in this area was a source of sand and volcanic deposits spread into a basin on its east ("inboard") flank of the arc, creating the Nooksack-Harrison unit (Figs.4-2, 4-3, 4-4).

This Nooksack-Harrison sedimentary and volcanic section (NH) inboard of the Insular superterrane is an important element of the Coast Plutonic Complex, extending the full north-south length of British Columbia, down even into the North Cascades (Figs. 4-2, 4-3). The Nooksack-Harrison unit, deposited along the east flank of the Insular belt far north of the present setting, moved south along a strike-slip fault (Fig.4-4, and further discussion later). Rock ages in the Nooksack-Harrison unit range from 235 to <114 Ma.

Other terranes making up the country rock to plutons within the CPC (Figs. 4-2, 4-3) are named differently in British Columbia and Washington across the Fraser River-Straight Creek fault, but can be correlated: Bridge River = Cogburn = Napeequa; Cadwallader = Chelan Mountains; Settler = Chiwaukum.

The Bridge River - Cogburn - Napeequa terrane is ocean crust. The Cadwallader - Chelan Mountains terrane consists of oceanic arc rocks that overlap in age and are closely associated with the Bridge River terrane, thus are envisaged to represent an arc developed on ocean crust (Fig. 4-5). The Settler-Chiwaukum terrane is likely a metamorphosed equivalent of the Methow basin forearc sediment, interpreted to be underlain by Bridge River and Cadwallader rocks (Fig. 4-5).

The Methow sedimentary beds, flanking Quesnellia on the west (Fig. 4-5), fall into older and younger packages. The older (Fig. 4-5 top, Fig. 4-6) consists of submarine fan deposits, ~160-100 Ma, which flooded off mountains to the east that were underlain by an active arc on the edge of Quesnellia. They were laid down in a basin floored by ocean crust and oceanic arc materials in front of the arc. This setting is termed a forearc basin. Importantly, these older beds mark the continental margin for that time.

At about 100 Ma everything changes. The forearc basin fills up, rises above sea level, and begins receiving stream-deposited, chert-rich, sediments from westerly mountains, as well as granitic sands from erosion of arc plutons in Quesnellia lying to the east (Fig. 4-5 bottom, Fig 4-6). This influx marks a major uplift of the westerly accreted terranes and thus marks the beginning of the Cascade Range and ancestral B.C. Coast Mountains.

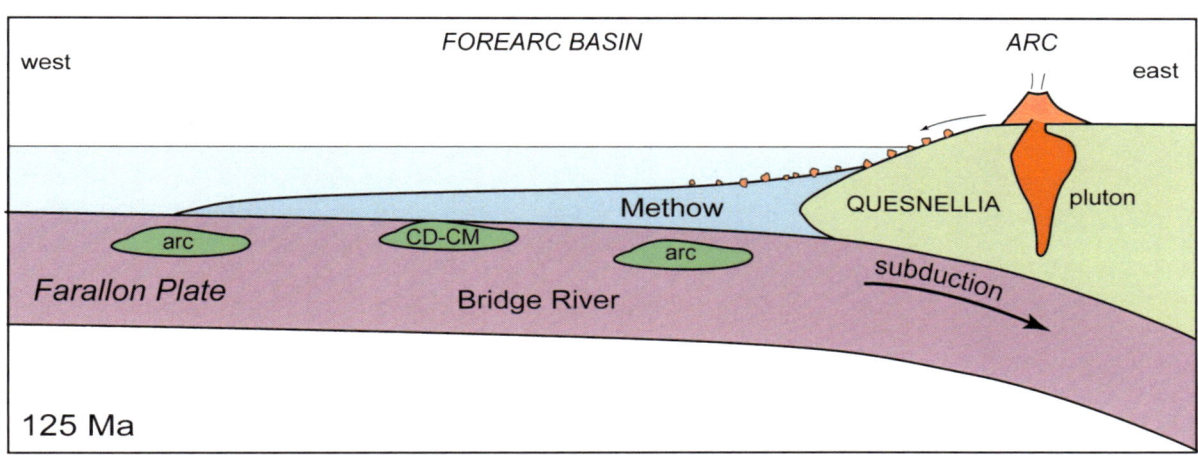

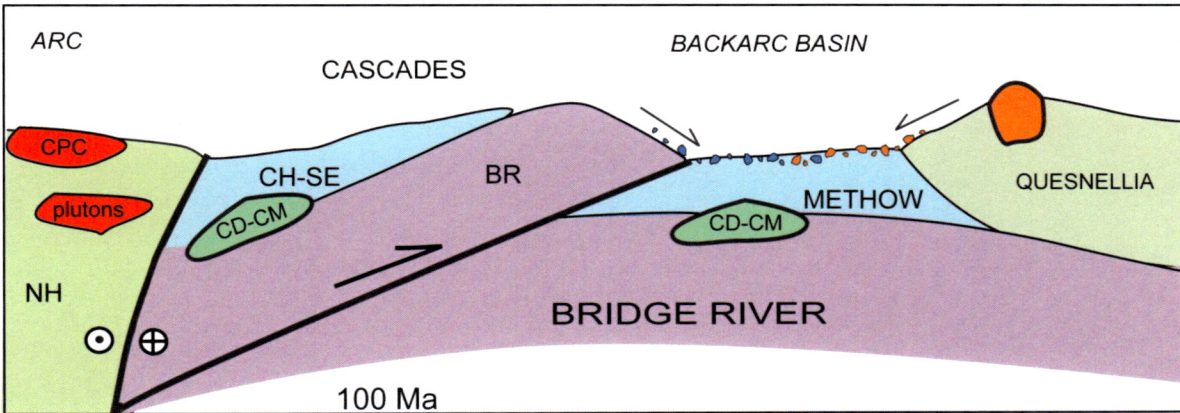

Fig. 4-5 Cartoon diagrams of tectonics and sedimentation in the Methow region.

Top: **Setting and origin of the older part of the Methow sedimentary sequence (160-100 Ma). Sand and mud are eroded from an active arc volcanic system on Quesnellia. Deposition occurs at the continental margin in a marine *forearc basin*. The arc is fed by subduction of the Farallon Plate.**

Bottom: **Arrival and collision of an outboard terrane, probably from the north (Fig. 4-8). The Farallon Plate is blocked off, and thus the volcano on Quesnellia goes extinct and is eroded down to the underlying pluton. The basin is uplifted, and the Cascade Mountains are initiated, including arc magmatism. Sediment comes from both sides into what is now a *backarc basin*.**

Fig. 4-6 *Above:* **View south from Slate Peak, looking down the Methow Valley, of early Cretaceous (140-100 Ma) forearc basin marine strata in the Methow area. The source was a magmatic arc to the east, on the edge of Quesnellia. The basin floor is ancient ocean crust of the Bridge River formation. At this time the Cascade mountains had not yet formed; thus, in the Methow area, open ocean lay to the west.**

Fig. 4-7 *Left:* **Chert-rich conglomerate, 100 Ma, in a backarc basin of the Methow area. Chert pebbles were carried by streams flowing from the uplifted Bridge River ocean crust to the west. This rock signals the time of origin of the Cascade mountains.**

Terranes of the San Juan Islands - northwest Cascades thrust system

Terranes in the thrust system (SJI-NWC, Figs. 4-2, 4-3, 4-4) have a different parentage than those of the CPC. They have correlations to rocks in Oregon and California, where incoming terranes were subducted in a series of stages beginning about 260 Ma, and carrying on until the Farallon Plate disappeared down the subduction zone in California ~ 30 Ma. Regional correlations are briefly described here. More detail about the rocks is in Chapter 5.

The basal nappe of the SJI-NWC is composed mainly of the Chilliwack Group, much of which is island arc volcanic rock. The formation also includes sedimentary rocks, some of which are related to Cordillera-long Paleozoic limestone beds, including those in Stikinia, Quesnellia, and similar terranes in Oregon and California.

An intriguing component of the Chilliwack terrane is the Yellow Aster Complex (Chapter 9), which contains fragments of a 450-400 Ma continental arc foreign to the North American Cordillera, but matching to the northern Appalachians and Greenland—a very long travel path.

Next up in the thrust sequence is the Bell Pass mélange, a scrambled assemblage of terrane fragments that includes a large chunk from the earth's mantle, the Twin Sisters dunite, as well as pieces of metamorphosed oceanic basalt and chert, and a relatively young unmetamorphosed sandstone.

Farther up in the thrust stack is the Easton terrane, an ocean-floor assemblage. Its depositional age is not yet known; the metamorphic age ranges from ~165 to 130 Ma. This rock does not have correlatives in the CPC, but does find close relatives in the Klamath Mountains and California Coast Mountains.

Finally, the uppermost nappe, a mixture of slices of the underlying nappes and other unrelated terranes, is termed the Mélange Belt. A large (miles across) part of this complex is a remarkably coherent ophiolite assemblage in the eastern San Juan Islands that is an intact, upturned, ocean arc exposed from mantle basement to sedimentary cover. This we call the Fidalgo Ophiolite. It correlates with ophiolites in the California Coast range and Klamath Mountains, but is not found in the CPC.

Notably, nearly all terranes of the SJI-NWC (except for the Mélange Belt) bear evidence of high pressure-low temperature (HP-LT) mineralogy, indicating subduction zone metamorphism. The terranes are separated by thrust faults that came after the subducted rocks were uplifted from the subduction zone.

The entire package of nappes lies in thrust contact over the south edge of the CPC (Figs. 4-3, 4-4). Exposure of the basal fault is seen in the Windy Pass thrust in the central Cascades, discovered by Bob Miller, and north of Mt Baker where the Chilliwack terrane is thrust over the Nooksack Formation, mapped by Peter Misch.

Big Slip of the Coast Plutonic Complex
A big shift to the south of terranes and plutons along the coastal area of southeast Alaska and northern British Columbia, is diagrammatically presented in Fig. 4-8. We surmise ~500 miles of displacement but have not yet found the fault. The hypothesis was first proposed by Jim Monger and others in 1994, and subsequently has had support from other geologists working in the region—notably George Gehrels and colleagues (Fig. 4-8).

What is it about the geology that suggests this proposed major dislocation? We need to look at the Methow region in northwestern Washington and its extension in southern British Columbia as viewed with restoration of the Straight Creek-Fraser River fault, shown in Fig. 4-3. The older arc plutons of the CPC form a double belt: one on the east side of the Methow basin in Quesnellia, and the other to the west in the main part of the CPC, including Wrangellia (Fig. 4-2). These two belts are interpreted to be disrupted parts of a once-continuous arc that was sliced and doubled up by along-coast sliding.

Next in evidence for the doubled up CPC hypothesis are depositional patterns in the Methow basin. The older sedimentary rocks of the Methow basin (Fig. 4-5) indicate a forearc setting, the arc lying along the western edge of Quesnellia. The sedimentation occurred in a deep marine basin with open ocean to the west, for a period from about 160 to 100 Ma at Washington latitudes. At the end of this time, the marine basin was gone—filled up with sediment. From this time forward, stream deposits from both eastern and western highlands piled up in the Methow region. The east-derived sediments are rich in granite sand and pebbles from the eroded Quesnellia arc. The west-derived sediments bear chert pebbles (Fig. 4-7) weathered from an ocean floor terrane—the uplifted Bridge River Group. The forearc Methow basin was apparently closed by about 100 Ma with arrival of the Bridge River terrane. The Bridge River had long since been accreted to the continental margin (>160 Ma), thus its appearance on the west side of the Methow basin is attributed to southward transit (Fig.4-8).

The Monger/Gehrels model (Fig. 4-8) has some interesting details that apply to our analysis of the southern CPC. The break-away fault in the north lies east of the older pluton belt which dates to 170-140 Ma. It also lies east of a back-arc basin in the northern Coast Mountains that bears sediment and volcanic deposits derived from the arc. This package of rock is part of, and rides along with, the displaced older arc. It travelled to our area of interest in the southern CPC, and is recognized as the Nooksack- Harrison terrane that we see in contact with the Methow basement assemblage (BR-CD) in the Harrison Lake area (Fig. 4-3). Minda Davis in Fig. 4-9 is looking down the general location of this proposed big fault.

No one has specifically identified outcrops of this coast-parallel CPC fault of huge displacement.

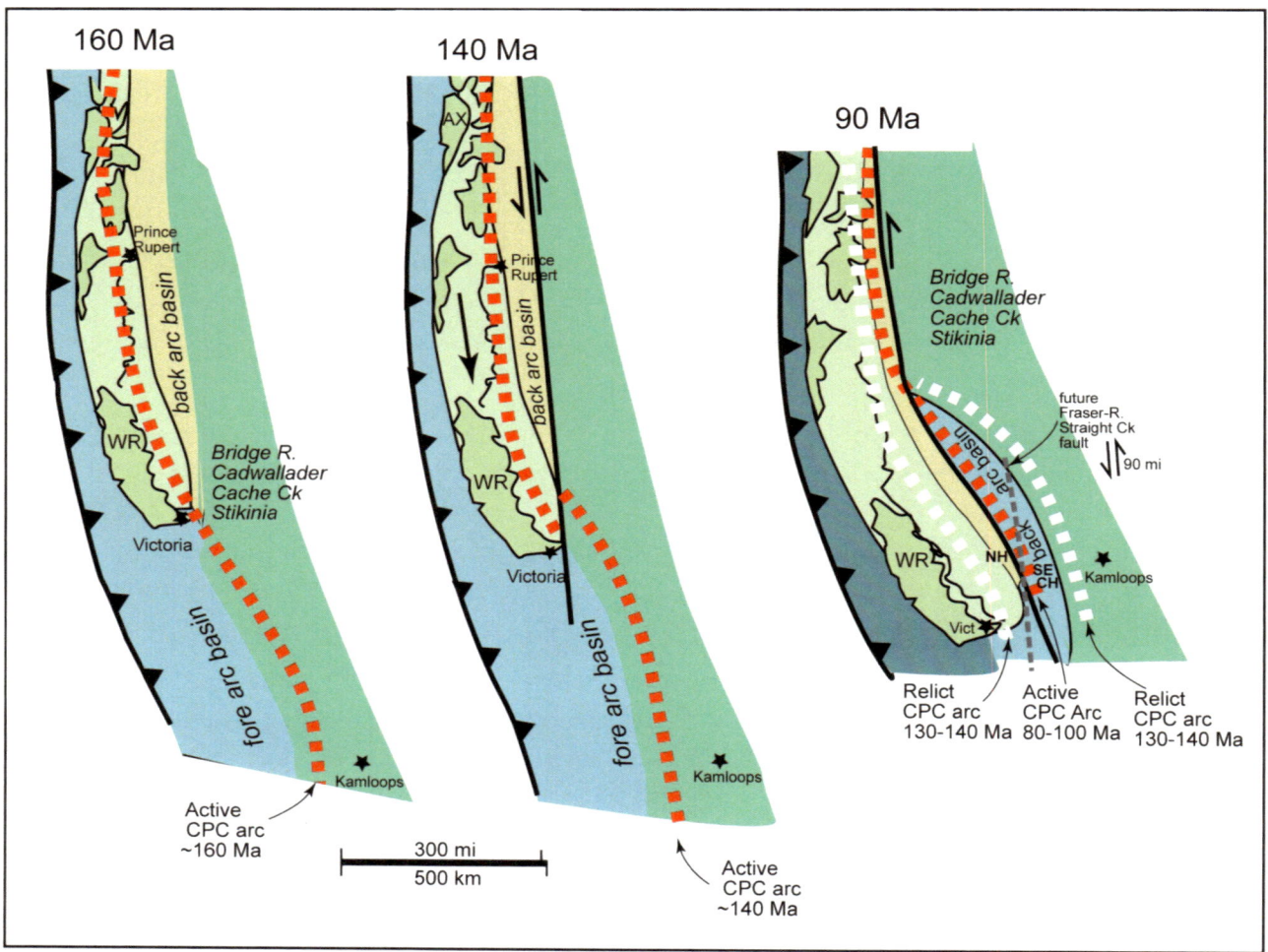

Fig. 4-8. Model for southward displacement of coastal terranes in northern British Columbia and southeast Alaska (diagram after Gehrels et al. 2009). The primary evidences are: the doubling up of the magmatic arc, extinction of the older arc in southern B.C., and the transition of the Methow basin from forearc to back arc, attributed to blockage of the incoming plate by the southward protuberance of the coastal terranes (Fig. 4-5).

Fig. 4-9 View north along Harrison Lake, British Columbia. Most of this country, along the lake and to the west, is Nooksack-Harrison terrane. Minda Davis is seen mapping the western edge of the Bridge River terrane. Thus the transition from NH to BR, where the hypothetical CPC fault runs, is in country somewhat east of the lake. Snowy Mount Breakenridge is seen on the right.

Looking in the general area of transition from the Nooksack-Harrison to Bridge River rock units in the Harrison Lake area, we see evidence of high strain and lineations consistent with the large strike-slip faulting (Figs. 4-10). Unfortunately for our model of sinistral strike-slip (west side moved south), the slip direction in these shear zones is clearly dextral. This finding does not support the model, but it doesn't kill it either. The sinistral displacement is older than ~100 Ma, and the dextral motion is younger than that age. So, maybe the shear zones we see today overprint an earlier movement direction. Or, the fault occurred but outcrops have not yet been found.

Regarding the sliding of terranes north and south along the coast, the Farallon plate reconstruction of Engebretson (Fig. 1-5) shows a coastwise *south* component of collision against North America before

110 Ma, and coastwise *north* after 110 Ma. This is pretty close agreement to timing of north and south terrane displacements discerned from geological observations.

Fig. 4-10. Outcrop of sheared NH terrane along the Big Silver River. A granite dike crossing the strain fabric gives a zircon age of 91 Ma, setting a younger limit to the age of deformation. Dan Gibson is examining structures, and thinking about bringing his Simon Fraser class here. Shear motion is strike-slip, west side north.

Mt. Baker is an active volcano in the Cascades; arc magmatism is ongoing, driven by subduction of the Juan de Fuca Plate (map location on Fig. 4-2). Mt. Baker serves as a time-lapse view back to the days of magmatism creating the Coast Plutonic Complex, the volcanic cover of which is long since eroded.

From this view on the north side of the volcano, we see two cones. The younger cone, on the left, has been built over the last ~40,000 years and is still active, with a vent in Sherman Crater on the south side of the summit that emits sulfur-laden steam more or less continuously. The old cone, on the right, named the Black Buttes, is just a shell of its former self. It was active 500-300 thousand years ago, then was breached by erosion. This volcanic complex is built on highlands underlain by the Nooksack Formation, the contact being mostly in the alpine meadows.

PART II – SAN JUAN ISLANDS - NORTHWEST CASCADES THRUST SYSTEM

CHAPTER 5

OVERVIEW

What are the tracts of rock that came from afar and assembled at our doorstep to form western Washington? How and when did they get here? What do these rocks tell us about growth of the North America Cordillera? A first step in this inquiry is a look at the stack of thrust sheets that underlie the San Juan Islands and extend into the northwest Cascades (Fig. 5-1).

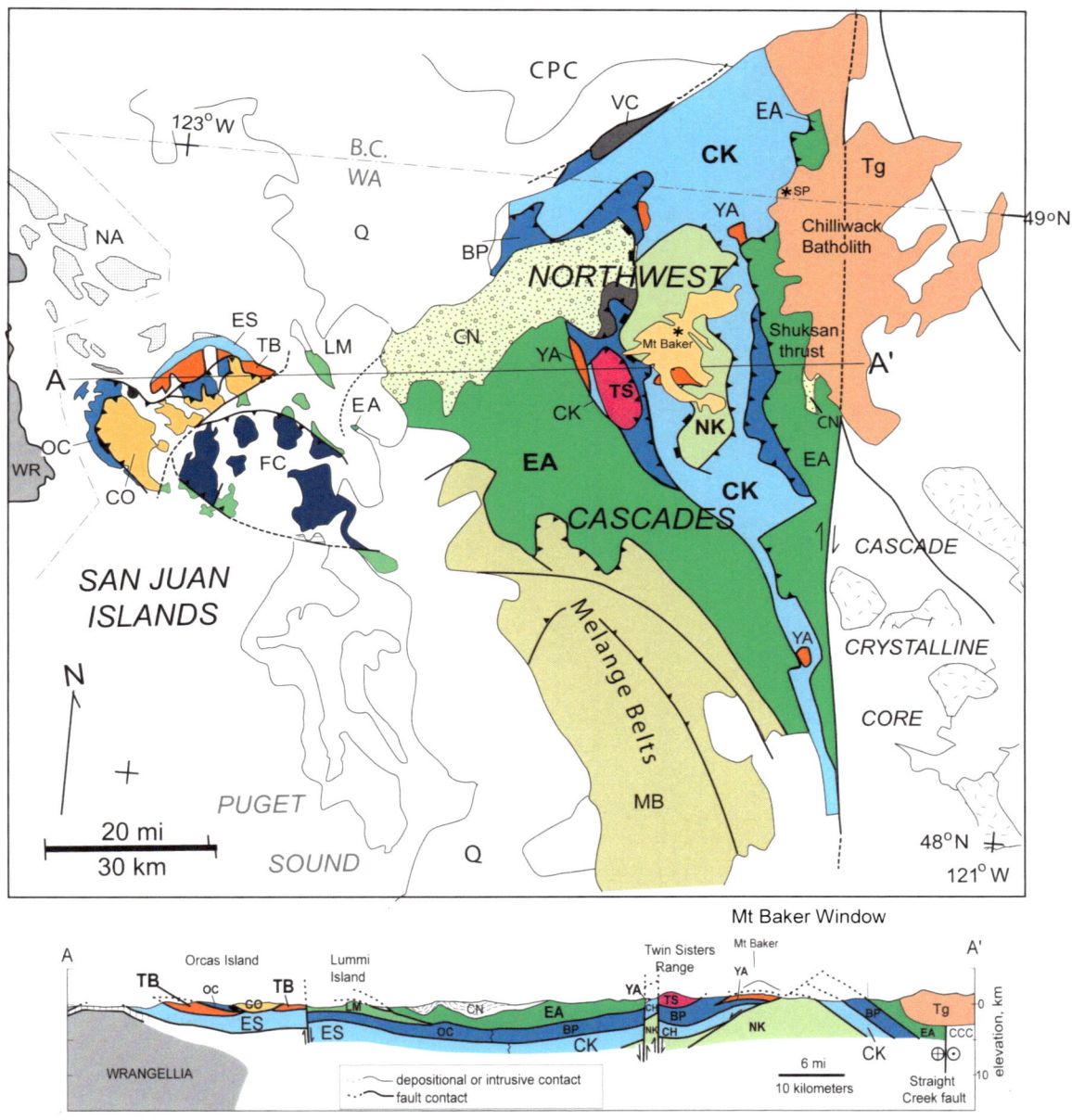

Fig. 5-1 Map and cross section of the San Juan Islands - northwest Cascades thrust system. BP=Bell Pass mélange, CN=Chuckanut Fm., CK=Chilliwack Gp., CO=Constitution Fm. CPC=Coast Plutonic Complex, EA=Easton Suite, ES=East Sound Fm., FC=Fidalgo Ophiolite, LM= Lummi Fm., MB=Mélange Belts, NK=Nooksack Fm., OC=Orcas Chert Fm., Q=Surficial deposits, SP=Slesse Peak, TB=Turtleback Cplx., Tg=Tertiary granite, TS=Twin Sisters, YA=Yellow Aster Complex.

When it comes to thrust systems, a common origin would be the underthrusting of sheets of rock by straight-on collision in an accretionary wedge (Fig. 1-2), or by other crustal contraction normal to the axis of the mountain chain, as in the Rockies or Alps. The San Juan Islands - northwest Cascade thrust system (SJI-NWC) (Figs. 5-1) is different in two ways: 1) the direction of thrusting is apparently parallel to the axis of the mountain belt, rather than perpendicular, and 2) the thrusts ride up and over the basement (obduction), not under the basement (subduction). The SJI-NWC is also unusual in the disparate origins and broadly-ranging ages of rocks comprising nappes (Fig. 5-2). Such geologic departures from the normal have challenged structural geologists and led to on-going controversy.

A critical exposure that reveals the nappe stacking order, and in fact the evidence that the faults separate nearly horizontal sheets, is seen on the hills flanking the valley of Baker Lake. Peter Misch, applying terminology gained in his early work in the Alps, describes this region as the "Mt Baker Window", because the nappe pile is eroded through a broad anticline and exposes the sequence of layers, faults, and their orientation (Fig. 5-1, 5-3).

From the bottom up, here is a brief description of the nappes:
The Chilliwack Group (including younger overlying rocks of the Cultus Formation) is dated by fossils and detrital zircons to range from ~ 400-175 Ma. This is a long-lived island arc sequence. Closely associated with the Chilliwack Group is the Yellow Aster Complex, a continental arc, yielding dramatic evidence for an origin in the northern Appalachians (Chapter 9)

The Bell Pass mélange, as the name implies, is a mixture, a phenomenally diverse assemblage consisting of: oceanic chert and ocean island basalt dated by fossils as 190-250 Ma, blueschist with a metamorphic age of 300-400 Ma, sandstone younger than 110 Ma zircon grains, Twin Sisters dunite from the earth's mantle, and Yellow Aster and Chilliwack rocks from the underlying nappe.

The Easton terrane is metamorphosed arc and oceanic crust. The metamorphic age ranges from 130 to 165 Ma. But the formation age of the arc and ocean crust is not yet known.

The Mélange Belt, as the name implies, is another structurally challenging package. The slices of rock include 150-180 Ma plutonic rocks, ocean floor basalt and chert, gneiss, mudstone/argillite, and sandstone important because it contains containing zircons as young as 74 Ma—by far the youngest rock of the entire nappe sequence.

The Nanaimo Group is made up primarily of

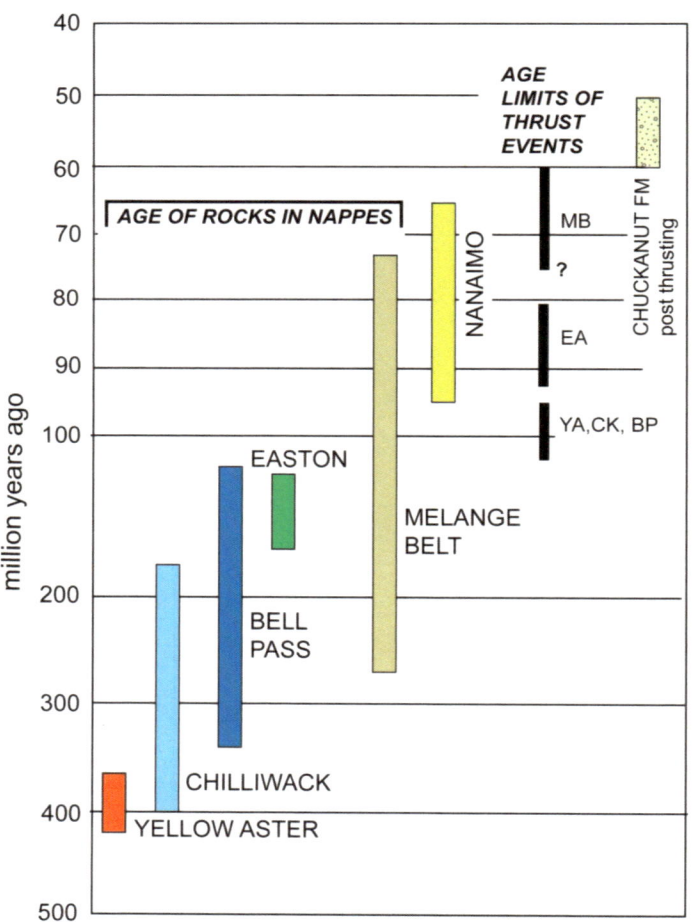

Fig. 5-2 Chart displaying age ranges of rocks in nappes; estimated age limits on thrusting events; the Nanaimo Group, synchronous with thrusting but mostly outside of the nappe assemblage; and the post-thrusting Chuckanut Formation.

marine sedimentary rocks. The sediments were deposited locally in a marine basin flanked by Wrangellia, the Coast Plutonic Complex, and the San Juan Islands thrust system. Some older parts of the Nanaimo occur within the thrust system, and although the structure cannot be discerned, it is probable that these parts of the Nanaimo were involved in the thrusting.

The Chuckanut Formation, dated from zircons and fossils to range from 60 -50 Ma, is deposited across the nappe structures, and thus gives us a minimum age for nappe emplacement.

How can we determine the ages of thrusting, as plotted on Fig. 5-2? The Chilliwack/Yellow Aster/Bell Pass nappe overlies the Nooksack Formation, and thus was emplaced after 114 Ma zircons deposited in the Nooksack sands, and also after 110 Ma zircons in the Bell Pass sandstone. This same nappe bears zircons distinctive of Yellow Aster rock (400-425 Ma) that are found shed into Nanaimo sandstone dated by fossils at 93 Ma. Thus we have age brackets on the Chilliwack/Yellow Aster/Bell Pass nappe emplacement as after 110 Ma and before 93 Ma.

The Easton Schist is probably correlative with rocks of the Ingalls Complex, the thrusting of which (Windy Pass thrust, Fig. 4-3) is dated at ~93 Ma, where the Ingalls rock cross-cuts older granite and is intruded by younger granites in the Mt. Stuart plutonic complex. A younger limiting age of the Easton emplacement is marked by chunks of this rock in conglomerates of the Nanaimo Group on Sucia Island in the San Juan Islands dated by fossils as 80 Ma.

The Mélange belt emplacement apparently postdates that of the Easton Schist in that fragments of the Easton are found in the Mélange. Another indicator is the young age, 74 Ma, of zircon–bearing sand in the Melange. A minimum age is that of the overlying, post-thrusting, Chuckanut Formation at 60 Ma.

An important feature of the nappes is that most of the formations (YA, CK, EA, BP) have components with high pressure-low temperature metamorphism that was inherited from subduction before nappe emplacement. This means that these nappes had a collisional history before being assembled in the SJI-NWC; they spent time subducted along the continental margin somewhere else.

The footwall (underlying basement) of the San Juan Islands - northwest Cascades thrust system is the Coast Plutonic Complex. This relationship is seen in two places: 1) at the Windy Pass thrust, mentioned above, in the central Cascades where the Ingalls complex is thrust over granites of the CPC (Fig. 4-3),

Fig. 5-3 Google Earth image of Baker Lake and surroundings, view north.
Cross section A-A' of Fig. 5-1 traverses the image, east-west, from about Twin Sisters to Bacon Peak. The section shows the geology at depth, illustrating the "Mt. Baker Window" labelled on Fig. 5-1 and revealing the nappe structure.

45

and 2) in the Mt. Baker Window, where the nappes are thrust over the Nooksack Formation that is country rock in the Coast Plutonic Complex (Chapters 4, 6).

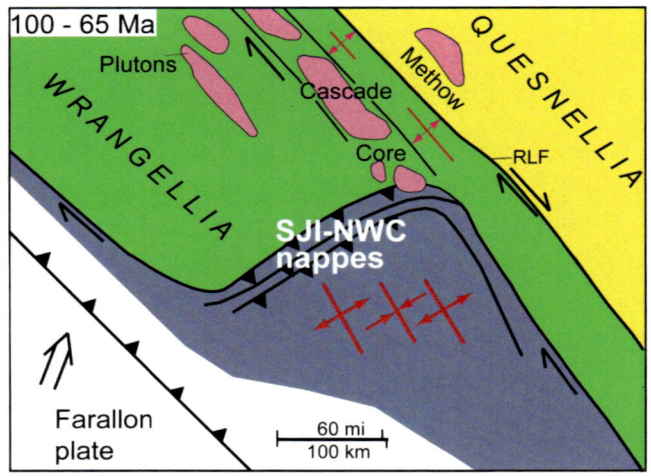

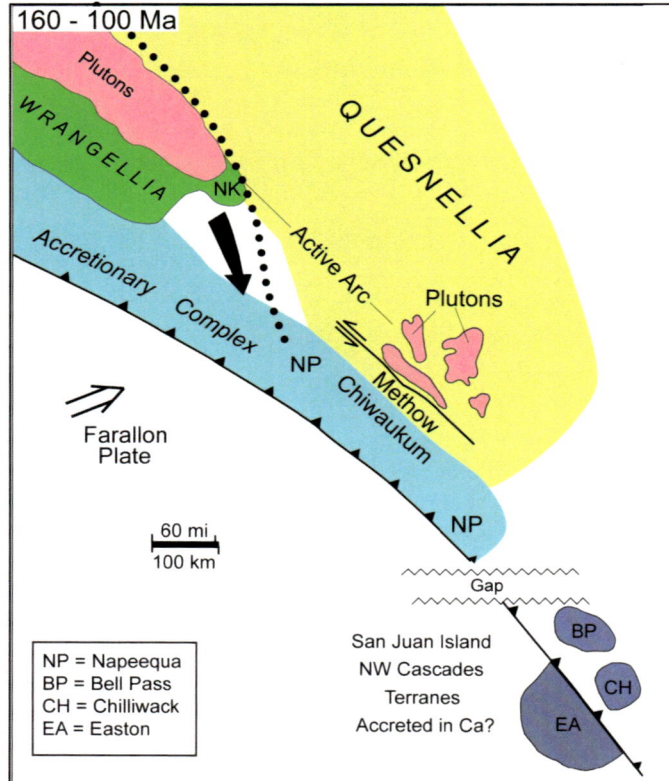

Fig. 5-4 Hypothetical sketches of the evolution of the San Juan Islands-northwest Cascades thrust system. Compare with Fig. 1-5. *Bottom:* **Terranes accreted along the continental margin in California. The Farallon Plate converged obliquely, sliding Wrangellia southward.** *Top:* **Later, convergence of the Farallon Plate shifted to a northward oblique component and brought the earlier accreted terranes northward into a collision zone at the south end of the Wrangellia-Coast Plutonic Complex corner area.**

In the San Juan Islands, the nappe basement is not observed, but is inferred to be Wrangellia based on the general sloping of nappes away from Vancouver Island. The Wrangellia connection is also supported by the apparent involvement of the older parts of the Nanaimo Group in the thrusting event (mentioned above); the sediments include material eroded from Wrangellia. These relations lead to the concept of an amalgamated Wrangellia-Coast Plutonic Complex as basement to the thrust system.

Next questions: which way did the nappes move, and from whence did they originate? An idea originated by Peter Misch (1966) and based on his work in the Alps and Himalayan orogens where mountain-building thrusts were driven out of a root zone between colliding continents (Fig. 2-2), is that the SJI-NWC nappes rose out of a collision zone on the east side of the Cascade Core and rode southwest over the Core to their present location. This process is orogen-normal contraction. In plate tectonic theory, the contraction could have occurred as Wrangellia moved during accretion against the continental margin, as suggested by subsequent workers, discussed in Chapter 12.

I have argued against this interpretation and in favor of an orogen-parallel trajectory (Fig. 5-4) based on 1) evidence that Wrangellia was already accreted by the time of nappe emplacement, 2) consideration that nappes riding over the Core, during pluton intrusions, would have been heated to the extent that the HP-LT minerals that we find would have been wiped out, recrystallized to higher T minerals, 3) the SJI-NWC would have been invaded by plutons of the underlying CPC, whereas none are found, 4) there are no vestiges of remnants of the SJI-NWC terranes in the suggested root zone on the east side of the Core, and 5) the nappe pile is, in fact, observed to lap onto the southeast edge of the CPC and Wrangellia (Fig.4-3)—the nappes moved northwest, an orogen-parallel displacement, not orogen-normal.

CHAPTER 6

NOOKSACK FORMATION

The Nooksack Formation is composed of an ancient volcanic arc and its associated sedimentary apron. It plays a key role in the drama of northwest Cascades thrusting—it is the footwall, i.e. the basement, over which the nappe pile rode (Fig. 4-3). It is mainly exposed in the Nooksack River Valley and runs south into the vicinity of Baker Lake (Figs. 6-1,6-7, 10-4).

The rock itself is not so exciting in outcrop compared to some of the over-thrust units. Two parts are recognized: an older igneous component termed the Wells Creek Volcanic Member, and an overlying clastic section of sandstone and shale.

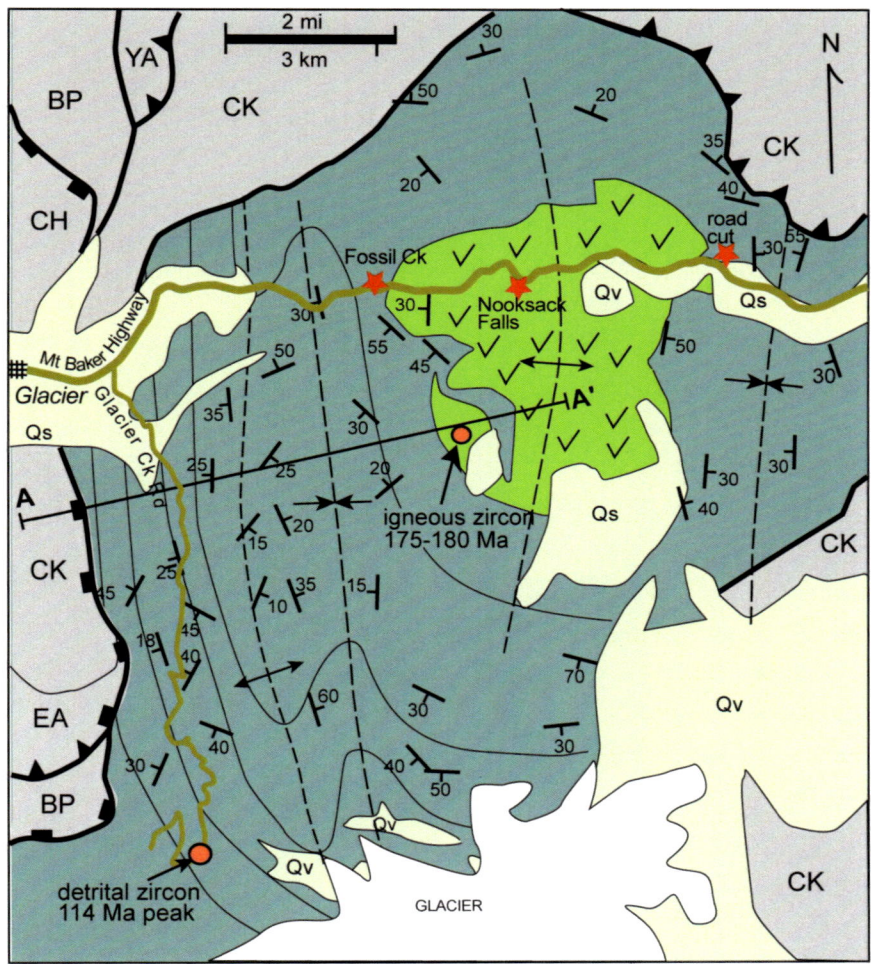

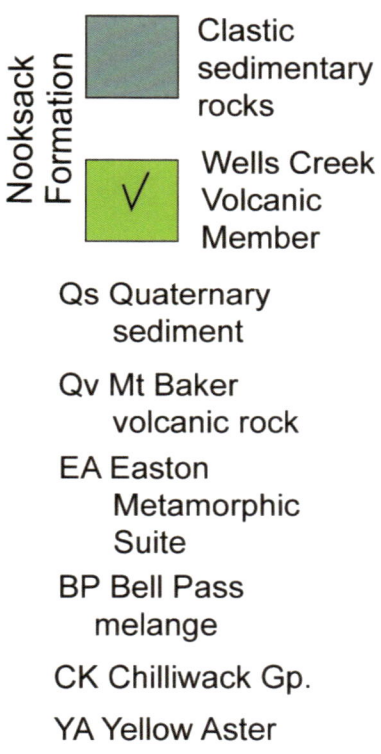

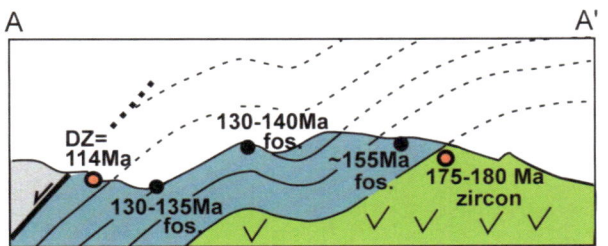

Fig. 6-1 Map and cross section of the Nooksack Formation and surroundings in the vicinity of the Mt. Baker Highway (located on regional map of Fig. 10-4). Strike and dip symbols show the orientation of bedding in the sedimentary layers. By these measurements, the fold structure and cross section are established. Zircon ages include detrital zircon sand grains "DZ" and igneous zircons in the volcanic rock. The "fos" refers to ages based on fossil identification.

47

My favorite outcrop of the Wells Creek Volcanics is in Nooksack Falls (Fig.6-2). The igneous rocks are fragmental and light colored, indicating explosive volcanism of silica-rich magma.

Fig. 6-2. Wells Creek volcanic rocks exposed in Nooksack Falls.

This fragmental pyroclastic rock exploded out of a volcano, mobilized by expanding gas. The rock composition is felsic, rich in quartz and feldspar. Technically, the rock is termed dacite.

Details of the fragmental rock texture are seen in the inset telephoto shot of an area on the cliff face indicated by a white arrow.

There is no safe access to these outcrops!

5 in

Overlying and inter-fingering with the volcanic rocks on a regional scale are sedimentary rocks: black shale derived from silt and mud, and interbeds of lighter colored sandstone. An eye-catching outcrop along the Mt. Baker Highway—"road cut" on map of Fig. 6-1— that I have enjoyed driving by count-less times on my way home from a mountain adventure, is a large cliff face that exposes interbedded sandstone and shale, crossed by dense fractures of slatey cleavage (Fig. 6-3). The intersection between bedding and cleavage marks the orientation of a large fold (Fig. 6-3 drawing).

The sand is composed of quartz, feldspar, and vol-canic grains.

Close-up of cleavage

Fig. 6-3. Shale and sandstone layers of the Nooksack Formation along Mt Baker Highway. Bed-ding layers are tilted ("dip") steeply to the right. Closely spaced slatey cleavage crosses bedding and dips steeply to the left. This geometry is typical of folded layered rock of this sort. Bedding defines the fold. Cleavage forms normal to the shortening direction by compression and alignment of tiny platy minerals, clay and mica, in the rock.

Fig. 6-4. Belemnite fossils in the Nooksack Formation, found in "Fossil Creek" along the Mt. Baker Highway (Fig. 6-1). The photo shows a rod-shaped calcite skeleton that supported a critter similar to a modern-day squid.

Fig. 6-5 *right:* Photomicrograph of sandstone collected from outcrop at the head of Glacier Creek road (Fig. 6-1). White grains are quartz; light green grains are altered feldspar; dark green grains are altered volcanic rock; and black grains are a metallic mineral—probably pyrite. This rock was crushed and processed for separation of zircon sand grains.

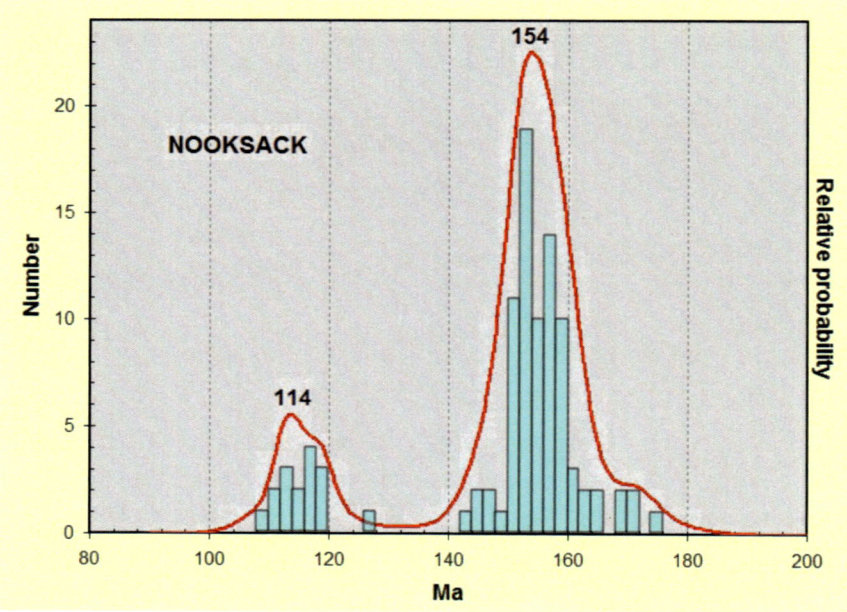

Fig. 6-6 *left:* Distribution of ages of zircon sand grains in the Nooksack Formation, photo above. Ages are in millions of years (Ma). The Y axis represents the number of individual laser spot analyses. The youngest population of zircons averaging at 114 Ma marks the oldest possible age of this sediment. And importantly, the 114 Ma age gives us a maximum age for the time of emplacement of the overlying SJI-NWC thrust complex (Figs. 4-3, 5-1).

What was the tectonic origin of the Nooksack Formation? The sediments are host to plentiful marine fossils and thus formed on the sea floor (Fig. 6-4). The underlying and inter-fingering Wells Creek Volcanics is also of marine setting, and of arc origin determined from mineralogy and chemistry. Thus, a best interpretation is that the Nooksack Formation was an oceanic island arc. Absence of very old zircons eroded from the ancestral North American continent (Fig. 6-6) supports the oceanic interpretation.

When we look at the evidence for age indicated by fossils and U-Pb in zircon, we see a range from ~175 Ma to 114 Ma. In the cross section (Fig. 6-1), the ages are progressively younger upward in the succession of rocks. From the variation of zircon ages (Fig. 6-6), we know that major volcanism was episodic; big activity from 150-160 Ma, and again but not so much at 114-120 Ma. So, sediments of the Nooksack Formation accumulated in a basin marginal to an island arc that at times was active, and at other times was quiescent but shedding much volcanic debris into a neighboring sedimentary basin. The total thickness of the unit was at least 2 to 3 miles (Fig. 6-1 cross section).

Let's look at the evidence for metamorphism of the Nooksack Formation. Several of the nappes that are thrust over the Nooksack contain aragonite and thus were metamorphosed at quite high pressure and low temperature (Fig. 3-15), in a subduction zone. Is this the story for the Nooksack itself? The answer is no. The diagnostic x-ray test for aragonite done for many Nooksack samples found none. Possibly aragonite formed but then reverted to calcite as pressure diminished during uplift. But the overlying aragonite-bearing nappes would have had a similar history and in them the aragonite is preserved. Metamorphic minerals found in the Nooksack include chlorite, pumpellyite, epidote, albite feldspar, and quartz; together these minerals indicate burial at depths as much as 5 to 6 miles, but not subduction zone depths. A further differentiation of the Nooksack from the nappes is that it is much less sheared up and structurally mixed than the nappes.

The Nooksack Formation has close relatives in terms of age, rock type and metamorphism in the Harrison Lake area of British Columbia. Others before have interpreted that the Nooksack Formation is a southern extension of that group of rocks, with which it is grouped here (NH, Figs. 4-2, 4-3).

Fig. 6-7. Google Earth image. View south of the landscape north of Mt. Baker. This terrain is largely underlain by the Nooksack Formation. Mt. Baker is perched on the Nooksack and other associated bedrock terranes that underlie the forested area. Baker Lake is in the distance to the southeast, and Twin Sisters to the southwest. DZ=detrital zircon dating sample, NF=Nooksack Falls, FC=Fossil Creek.

51

CHAPTER 7
BELL PASS MÉLANGE

In geology the term mélange, as in ordinary usage, refers to a mixture. This is a mixture of chunks of rocks of all sizes, shapes, lithology, and origin, without a sense of internal structure other than shear zones. When I first started thinking about geology of the northwest Cascades, I had never seen such a mixed up assemblage of rocks as in this area, and I thought of the whole region as a huge mélange. This outlook amounts to giving up on sorting out separate units and mapping structure. Peter Misch didn't call rocks of the entire northwest Cascades a mélange; he mapped the internal units, and was able to make sense of the geology as a stack of thrust sheets.

So, after some explorations in the mountains, I got with this program. But in the process of sorting out the mappable units based on rock type, degree of metamorphism, and age, there still remained some zones of rock intermixing that defied regional mapping. Ralph Haugerud, in his M.S. thesis (1980), and Rowland Tabor and others (2003), brought the term mélange into play, referring to a thick sheet in the regional thrust stack that has such a complex mixture of unrelated rocks, down to the outcrop scale, that regional mapping is precluded. This rock package Tabor and others termed the Bell Pass mélange, named for a locale just east of the Twin Sisters, where the regional mapper would have to give up in trying to show all the details (Fig. 7-1 next page).

Here is a list of the main units of the Bell Pass mélange, mostly following Tabor and others, 2003:
—Yellow Aster Complex
—Chert-basalt assemblage
—Baker Lake blueschist
—Vedder Complex
—Elbow Lake sandstone
—Twin Sisters dunite and other ultramafic rocks

The Yellow Aster Complex (Chapter 9) we associate with the Chilliwack terrane, and ascribe its presence in the Bell Pass mélange to have occurred by entrainment during inter-slicing of rocks as part of the Bell Pass emplacement event.

The Chert-basalt unit is an assemblage of ribbon cherts (Fig. 7-2) and ocean-island basalts, very much sheared up but extensive throughout the mélange. Radiolarian fossils in the chert give ages ranging from about 300 to 200 Ma (Permian - Triassic). Aragonite is present,

Fig. 7-2. Ribbon chert along the east shore of Baker Lake. This rock, part of the Bell Pass mélange, is formed by deep sea settling of plankton and clay-rich ooze.

52

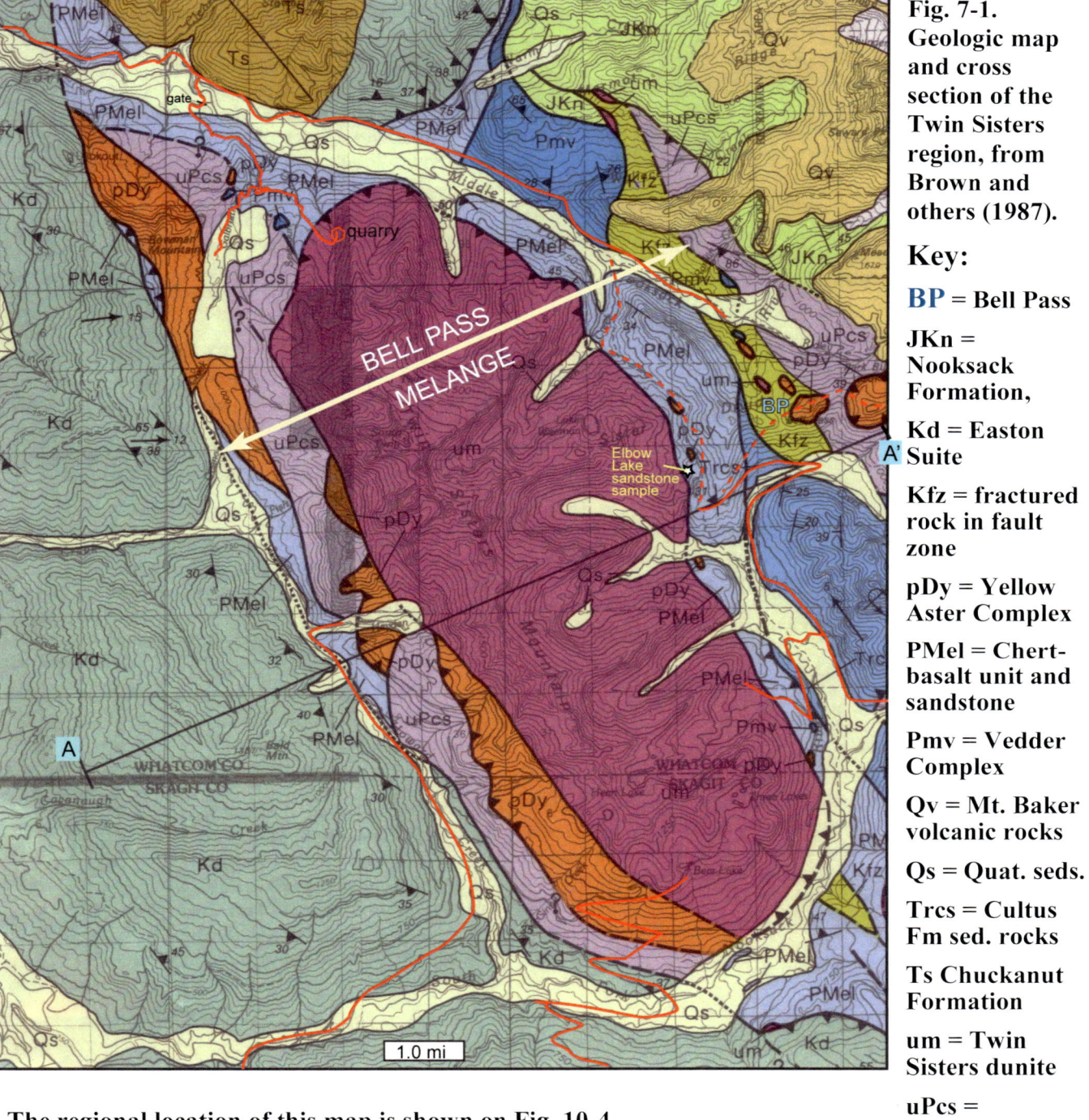

Fig. 7-1. Geologic map and cross section of the Twin Sisters region, from Brown and others (1987).

Key:

BP = Bell Pass

JKn = Nooksack Formation,

Kd = Easton Suite

Kfz = fractured rock in fault zone

pDy = Yellow Aster Complex

PMel = Chert-basalt unit and sandstone

Pmv = Vedder Complex

Qv = Mt. Baker volcanic rocks

Qs = Quat. seds.

Trcs = Cultus Fm sed. rocks

Ts Chuckanut Formation

um = Twin Sisters dunite

uPcs = Chilliwack Gp.

The regional location of this map is shown on Fig. 10-4.

indicating subduction zone metamorphism. The Chert-basalt unit correlates with the "Orcas Chert" of the San Juan Islands (Fig. 5-1), that bears fossils of the Tethyan sea—from south Asia. The Chert-basalt unit together with the Orcas Chert constitute a very far-travelled terrane and a thrust sheet of regional extent.

The Baker Lake Blueschist is a small sliver of greenschist and blueschist metamorphic rock at the head of Baker Lake. The metamorphic age is similar to that of the Easton Suite (130 Ma), but the original rock was mainly ocean-island basalt, not ocean-ridge basalt as in the Easton.

The Vedder Complex is a unit of sporadic regional extent, occurring in patches in the Cascades, and in the San Juan Islands, where it is named the Garrison Schist. The most extensive component of the Vedder Complex is relatively high-grade blueschist metamorphic rock in the form of amphibolite and quartzite—-another deep ocean protolith of basalt and chert. Measured K/Ar and Rb/Sr ages of metamorphism range from ~200-300 Ma, marking an older event of subduction than that of the Easton Suite. In both the Cascades and San Juan Islands, this rock is closely associated with the regional Chert-basalt unit and Orcas Chert, that are much less metamorphosed but have the same age and type of protolith. This rock unit is described in more detail by Gibson and Monger (2014, online access given in references).

A sandstone unit mapped by Dave Blackwell in the vicinity of Elbow Lake (Fig. 7-1) has particular significance in understanding the Bell Pass mélange. This rock has a decent population of detrital zircons at about 110 Ma—giving us a maximum depositional age (Fig. 7-3). Three younger zircons,

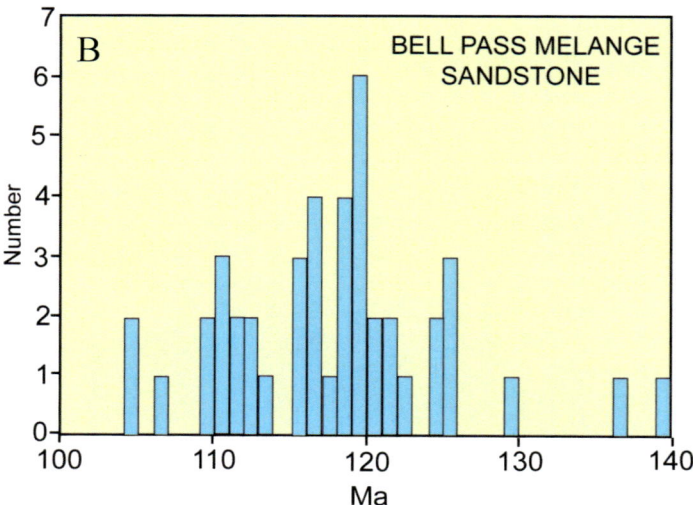

Fig. 7-3. Sandstone in the mélange at Elbow Lk (located on Fig. 7-1). A) sandstone texture: poorly sorted, angular rock fragments and sand, B) detrital zircon age pattern indicating a most likely maximum depositional age of 110 Ma. C) Liz Schermer collecting on a steep, slippery outcrop above Elbow Lake.

54

down to 105 Ma, suggest an even younger age, but could reflect machine instability or mineral alteration. In contrast to other rock units in the mélange, the sandstone lacks high-pressure metamorphic minerals and shows virtually no metamorphic fabric. The implication is that the HP-LT metamorphism found in other components of the mélange was acquired before the mélange was assembled. And further, that thrusting of the Bell Pass mélange occurred after 110 Ma.

Twin Sisters dunite

The Twin Sisters range is underlain by one of the most intriguing rock units of the Cascades—a giant slab of fresh dunite derived from the earth's mantle (Figs. 7-4 to 7-10). In outcrop, we see a body some ten miles long and three miles wide. From aerial gravity and magnetic surveys, we know it is a relatively thin slab, only about a mile thick (Fig. 7-4). Structurally, this slab lies in the plane of regional

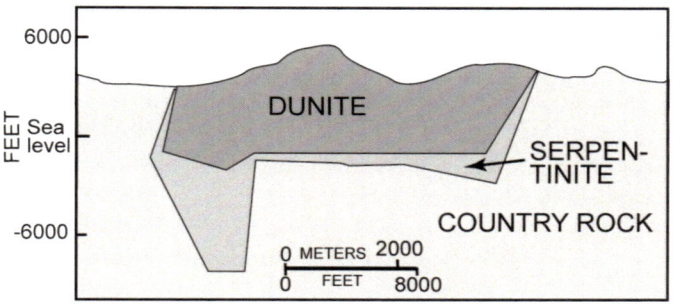

Fig. 7-4. Schematic depth profile of the Twin Sisters dunite, based on gravitational and aeromagnetic measurements that distinguish the extent at depth of the three rock types: dunite, serpentine, and country rock. From data plot of Thompson and Robinson, 1975; in reality, rock boundaries are rounded.

thrusting of the Bell Pass mélange. By some inscrutable process, the mechanism of mélange formation grabbed mantle rock and brought it to the surface.

The dunite of the Twin Sisters is mostly made up of olivine and a little chromite (Figs. 7-6, 7-7, 7-8). Pyroxene is locally an accompanying mineral —most is dark grey orthopyroxene (enstatite), but greenish clinopyroxene (diopside) can be found also. The Twin Sisters dunite is a metamorphic rock. It has a fabric defined by trains of chromite crystals. In microscope view (Fig. 7-9), the olivine grains look metamorphic; they lack crystal faces, are grown together, and are slightly elongated, defining a metamorphic plane of foliation.

Fig. 7-5. The Twin Sisters range, view west. The rock visible is nearly all dunite. Olivine comprises >90% of the rock, chromite and pyroxene making up the remainder. Other such large bodies of olivine-pyroxene rock on earth are mostly altered to serpentinite, but not the Twin Sisters which is reckoned to be the largest mass of fresh dunite on earth.

Fig. 7-6. Freshly broken, green-looking dunite, with a scattering of black chromite grains. Incipient weathering alteration on the surface of the sample is iron stain, seen everywhere on outcrops of the dunite except on recently broken surfaces.

Fig. 7-7. Yellow-stained dunite with chromite bands that define a planar fabric. This fabric is cross-cut by a mass of grey pyroxene, under my finger. The dunite fabric, planar in 3D, is metamorphic, inherited from solid-state ductile flow in the earth's mantle. The grey pyroxene rock can be interpreted as an igneous rock crystallized from magma developed by partial melting of the dunite.

Fig. 7-8. Close-up of the thick chromite band in the above photo. Width of view is about one inch.

Locally, pyroxene masses cut across the metamorphic fabric of the dunite, suggesting a magmatic origin (Fig. 7-7). As discussed in chapters 1 and 11, pyroxene in mantle rock over subduction zones melts under certain conditions, leading eventually to basalt magma that rises upward feeding arcs. Could this initiation of magma in the Twin Sisters body be related to the feeding of an ancient arc?

The best access to the Twin Sisters is via the road up the Middle Fork of the Nooksack River (Fig. 7-10). Both the quarry and the alpine region accessed from Dailey Prairie present excellent exposures. (I once slept a summer night on the summit ridge of the North Twin—fabulous).

Fig. 7-9. Microscope view of Twin Sisters dunite. Olivine grains are bright colors in crossed polarized light. The texture is metamorphic, termed granoblastic.

Fig. 7-10. *above:* **Google Earth image of the Twin Sisters range—dunite outcrop is marked by the yellow color, where vegetation is absent.**

Right: **Route to the olivine quarry and Dailey Prairie. From the Mosquito Lake road, turn off on Forest Service road 38 into the valley of the Middle Fork of the Nooksack River. Go 4.8 miles, then turn onto a side road and across a bridge over the river to a nearby gate. Then ~3miles on foot to the quarry, or to Dailey Prairie at base of the peaks.**

CHAPTER 8
CHILLIWACK GROUP &
CULTUS FORMATION

A long stretch of geologic time, from ~400 to 170 million years ago, is represented in the North Cascades by marine sedimentary and volcanic rocks of the Chilliwack Group and the overlying Cultus Formation (Figs. 8-1 through 8-10). The volcanic rocks are of island arc origin, and the associated sedimentary rocks formed largely by accumulation of silt and sand eroded from the arcs. Limestone is also part of the sedimentary assemblage, formed mainly as shellfish and reef accumulations marginal to the oceanic volcanoes. Remember from Chapter 1 that modern day examples of such island arcs are abundant in the western Pacific Ocean (Fig. 1-4).

Where was this arc? In pursuing this question, we benefit from the finding that the Chilliwack Group shares fossil types, termed the McCloud assemblage, with broad tracts of similar island arc rocks in British Columbia to the north, and in northern California to the south (Fig. 4-1).The fossil assemblage

Fig 8-1. Mountainous terrane carved into the Chilliwack Group. In the distance on the left is Canadian Border Peak, then American Border Peak (top in the cloud); the dominant peak is Mount Larrabee, and far right are the Pleiades. Thrust faults are seen high on the right side of Larrabee where grey gneiss of the Yellow Aster Complex is sliced into reddish volcanic rock of the Chilliwack Group. The Pleiades are Yellow Aster rock. A trail from the Twin Lakes road gives access along the meadow to the basin at the foot of Larrabee.

provides some clues. Crinoids, corals, and fusulinids (small pod-shaped, bottom-dwelling shellfish) suggest affinity to the continent—that is, the sediments formed in the ocean but were within the realm where animals of continental seas could mix in.

But there is still ambiguity: What part of North America was close by when the Chilliwack sediments were deposited? The Chilliwack sediments lay on the ocean floor for hundreds of million years, during which the oceanic plates were on the move. Recent data on zircon ages find affinity of the 400 Ma Yellow Aster Complex (Chapter 9) to the northern Appalachian mountains. We think (but are not sure) that the Yellow Aster Complex is connected to the Chilliwack Group. A hypothesis for the Chilliwack is that it formed as an island arc, more or less continuously active, while it was carried some thousands of miles by plate tectonics through the ancestral Pacific Ocean, possibly also through the arctic realm, and down the west flanks of Laurentia. We'll see more about this in the next chapter.

In addition to rock exposures illustrated here, the Chilliwack Valley just across the border in British Columbia offers excellent roadside outcrops of the Chilliwack Group, as well as Cultus Formation, and Vedder Complex. A recently published field guide to this area is available online—see Gibson and Monger (2014) in the *references*. This field guide was written for the geologist, but much is there also for the layman.

Fig. 8-2 Volcanic rocks in the Chilliwack Group.

Top: **Igneous breccia on Sauk Mountain, near the parking lot. Andesitic igneous rock is blasted into fragments by volcanic eruption. The green color comes from crystallization of low-grade metamorphic minerals, especially chlorite, during tectonic burial.**

Bottom: **Pillow basalt at Mt. Ann. Rounded pillows form as streams of basalt magma pour into the sea water.**

Fig. 8-3. Chilliwack sandstone in the Red Mountain limestone quarry. The larger grains appear to be vein quartz—possibly from a metamorphic rock; if that's true, an orogenic zone was nearby. More study of this rock is needed.

Fig. 8-4 Histogram of detrital zircon ages from the sandstone in Fig. 8-3. Maximum depositional age is about 370 Ma. The source area of the sand shows a broad spread of ages, going back to 410 Ma.

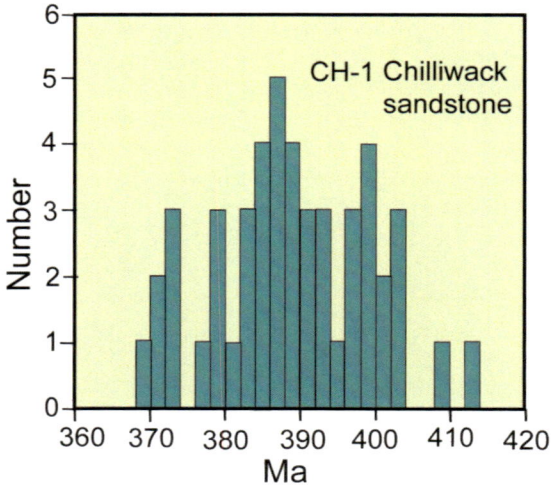

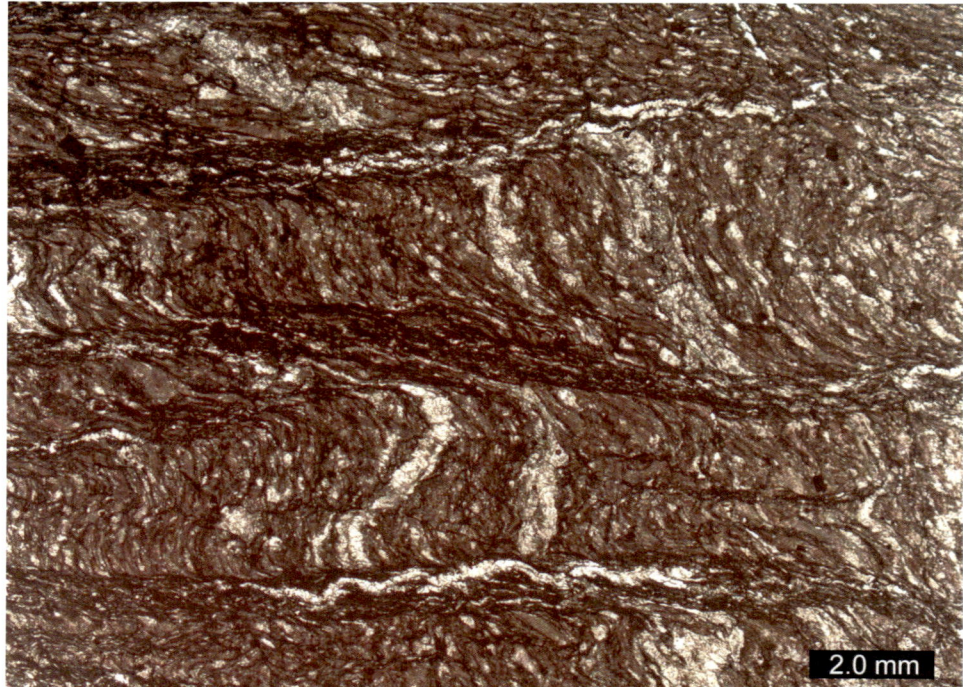

Fig. 8-5. Metamorphosed siltstone of the Chilliwack Group. Deep burial, compaction, and shearing affected this rock. Folded lighter-colored layers are original sedimentary beds rich in quartz and feldspar silt grains. Darker rock is clay-rich silt with individual quartz and feldspar grains discernable. Two phases of deformation are recognizable. The older phase is marked by flattened grains, defining a steep foliation. The younger deformation has folded the earlier foliation, and created nearly horizontal zones of new foliation that truncate the first foliation. Metamorphic minerals (including aragonite) in other rocks of the Chilliwack Group indicate high-pressure metamorphism in a subduction zone. The metamorphic fabrics in this sample likely formed by the subduction process, but we can't say much more than that, as for example the direction of subduction.

Fig. 8-6 *above:* Fossiliferous limestone from a quarry near the town of Concrete, Washington. The fossils are parts of a crinoid stem (Fig. 8-8), dating this part of the Chilliwack Group as Early Pennsylvanian: ~ 320 Ma.

Fig. 8-7 *right:* Relic of the past: cement plant in Concrete, WA, fed by Chilliwack limestone from the nearby quarry, decommissioned in 1968.

Fig. 8-8 *above*: Structure of a crinoid. The fossils from the Chilliwack Group illustrated in Fig. 8-6 are an assemblage of "columnals", segments of the stem, and thus are only a small part of the original animal. The crinoid as a whole is anchored to the sea floor. Crinoid relatives in the modern world include starfish and sand dollars.

Reference: http://dcfossils.org/index.php/gallery9/

Fig. 8-9.

Well-polished limestone outcrop of the Chilliwack Group along the east shore of Baker Lake.

Fig. 8-10 *Top:* View north of Canadian and American border peaks, from Gold Run Pass. The mountains are composed of arc-related sedimentary and volcanic rocks of the Chilliwack Group. Black arrow marks the campsite.

Bottom: Dark andesitic to basaltic rocks are exposed at the col on the north edge of American Border Peak. Brooke Sandahl emerges after a stormy night—-the tent flattened several times.

CHAPTER 9
YELLOW ASTER COMPLEX

From start to finish in looking at the Yellow Aster Complex, it's a "weird duck" in the context of the northwest Cascades thrust system. There are basically two components: metamorphosed quartz-rich sandstone and granitic intrusions. The sandstone occurs as two varieties: one is an arkose rich in quartz, with lesser amounts of feldspar, and the other is a quartzose sandstone with an appreciable limy component and interbeds of marble (Fig. 9-1). Zircon age patterns suggest mutually different provenances for their origin (more later). The ringer here in terms of compatibility with neighboring rocks is that quartz sand indicates a stable, low-relief continental provenance, a plate tectonic "passive margin" (the trailing edge of a plate on the move), where chemical weathering removes all minerals except quartz. Sandstones in other rocks of the SJI-NWC are rich in rock fragments, including: basalt, ribbon chert, and subduction zone metamorphic fragments. These features indicate origin in an accretionary complex, or an ocean arc, or the sea floor—a collisional margin.

Besides the metamorphosed sandstone, the Yellow Aster Complex is notable for an array of igneous intrusions (Fig. 9-1), mostly granitic, but ranging to basalts and gabbros. These rocks apparently represent the beginnings of an arc intruded into the continental margin.

Fig. 9-1. Outcrop of Yellow Aster Complex in Schreibers Meadows on the south side of Mt. Baker. This folded, layered rock is metamorphosed quartz-rich limy sandstone (calc-silicate), now metamorphosed to gneiss. Zircons from the sandstone are mostly 400 Ma in age. The white intrusive rock is granite of an age closely similar to the sandstone. To reconstruct the geologic environment, we envisage a near-shore marine sedimentary basin, with calcareous shellfish, receiving chemically mature sand from a low-relief landmass. Folding and arc-related intrusions happened shortly after sediments were laid down.

63

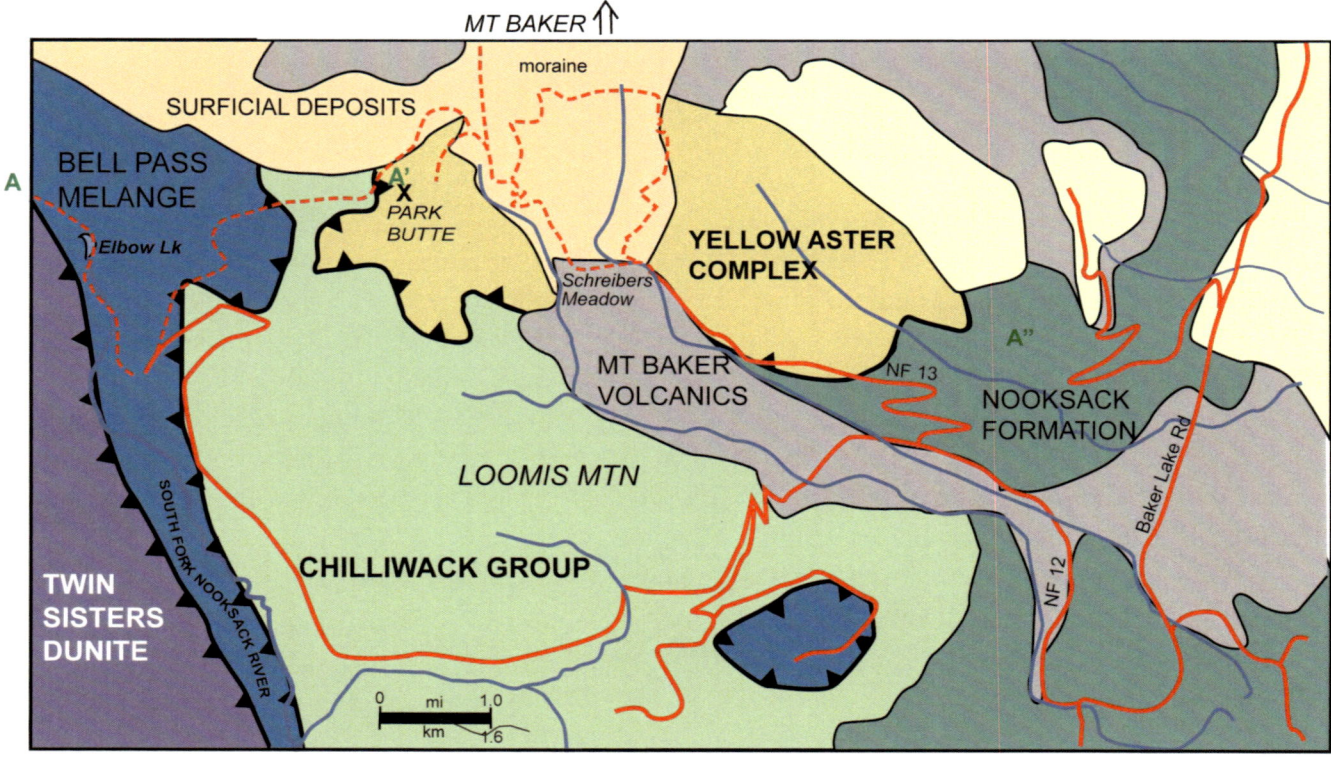

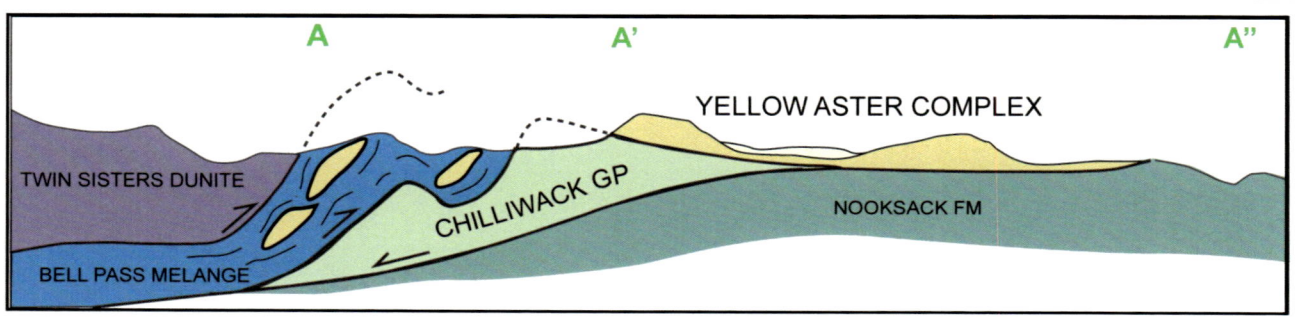

Fig. 9-2. *above:* **Map and cross section of the structural assemblage of Yellow Aster Complex in the SJI-NWC thrust system along the south side of Mt. Baker in the Schreibers Meadow area. Regional location on Fig. 10-4.**

Fortunately for hikers, the Yellow Aster Complex is well exposed in two alpine regions with good trail access: Schreibers Meadows and Yellow Aster Meadows (Figs. 9-2 to 9-6).

The Yellow Aster Complex does not occur as a contiguous mass over a broad region. Instead, we find slabs of the rock, up to miles across, sliced in between nappes of the Chilliwack Group, Easton Suite, and Bell Pass Melange. Small lenses and blocks of Yellow Aster Complex are a common component of the Bell Pass Melange.

Fig. 9-3 *left:* **Cliff and talus pile of Yellow Aster Complex at the west edge of Schreibers Meadows, as labelled on Fig. 9-2 above. The talus is very accessible and reveals excellent Yellow Aster exposures, as in the rock shown in Fig. 9-1.**

64

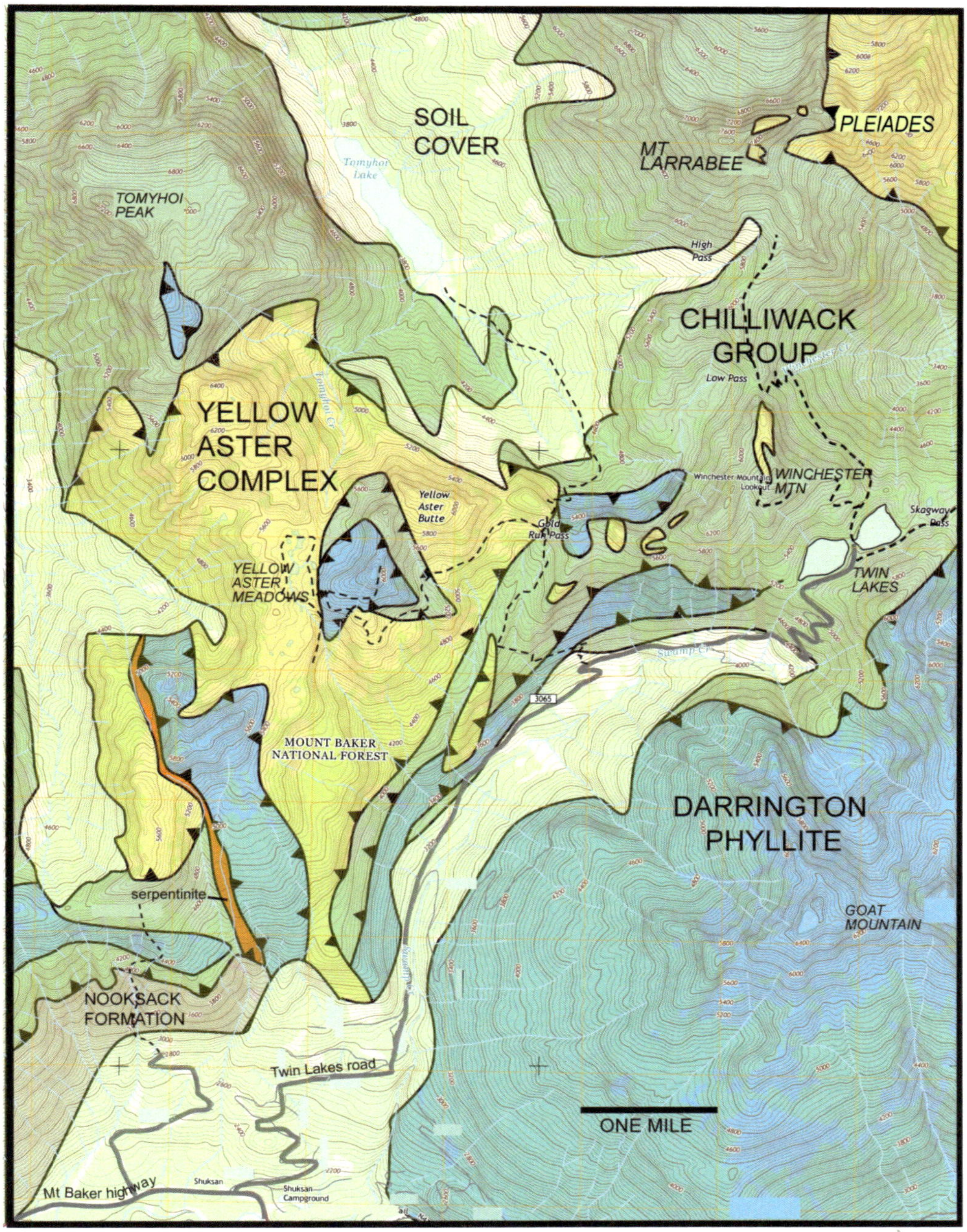

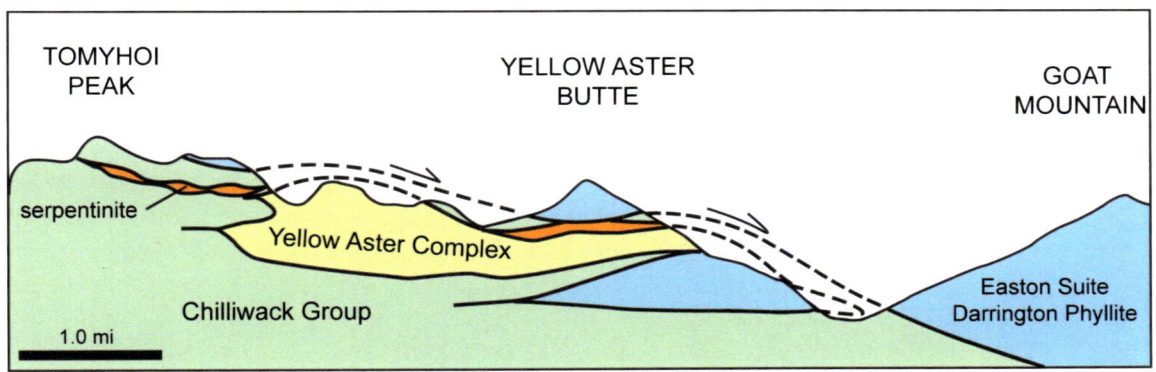

Fig. 9-4. *above:* Geologic map of the Twin Lakes and Yellow Aster Meadows regions, both accessed from the Mt. Baker Highway and Twin Lakes road, and beyond by excellent trail systems. Location on the regional map of Fig. 10-4. *below:* Cross section from Goat Mountain to Tomyhoi Peak.

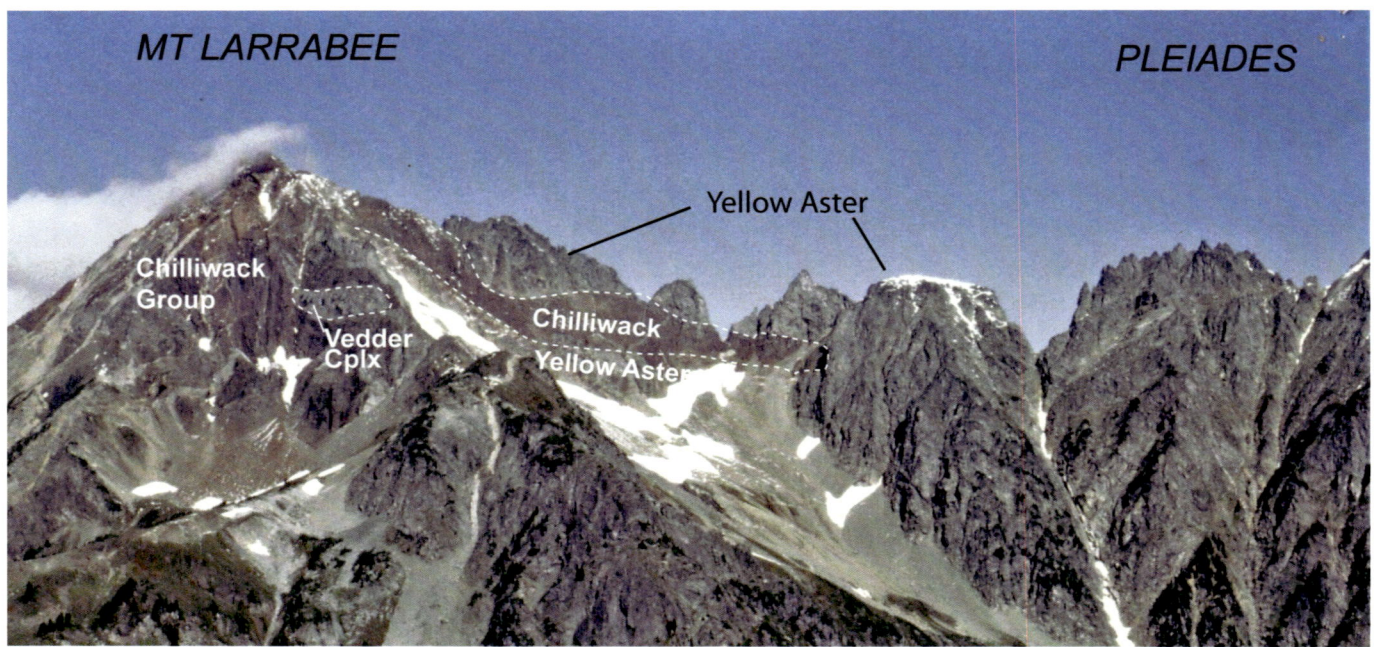

Fig. 9-5. View of interleaving of the Yellow Aster Complex and Chilliwack Group in the vicinity of Mt. Larrabee and the Pleiades. The slice of Vedder Complex marked on Mt. Larrabee is not a relative of either of the other two rock units—an indication of relatively profound structural mixing.

Fig. 9-6. Google earth image of the Yellow Asters Meadows - Mt. Larrabee - Pleiades area.

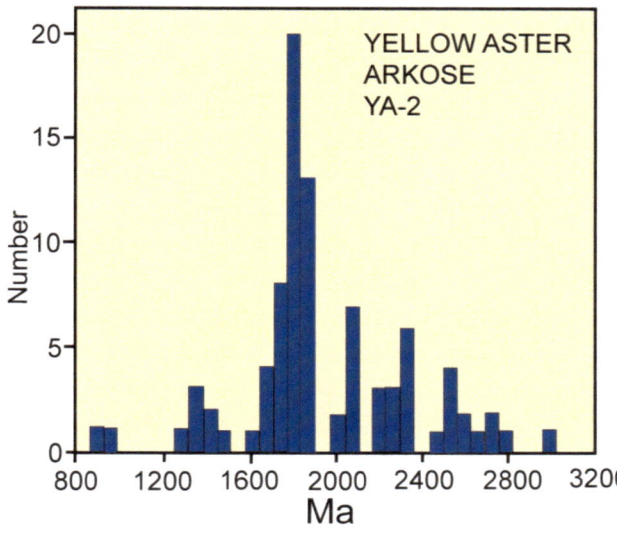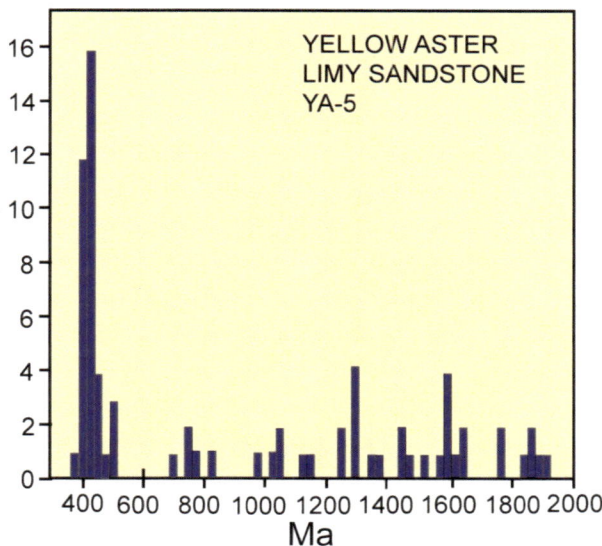

Fig. 9-7. Ages of zircon grains in metamorphosed sandstones of the Yellow Aster Complex. The two samples have quite different age populations, representing different source areas. Sand in the arkose is from northwest Laurentia. The limy sand likely came from eastern Greenland.

Zircon ages are greatly valuable in unravelling the history of the Yellow Aster Complex. Early age determinations were made on bulk zircon samples, which for sandstones is not so useful, because the individual zircon sand grains can have ages spread over 100s of millions of years. The first detrital zircon study was done by George Gehrels and me in 2007, investigating an arkose (left graph, Fig, 9-7). The ages are all Precambrian, spread from 1000 to 2800 Ma, with a good peak at 1800 Ma. This pattern fits well with the zircon age signature of sediments derived from Precambrian rocks of the northwestern part of the ancient craton Laurentia (Fig. 4-1). So, there is no evidence of long-distance terrane travel; the sandstones formed along the continental margin somewhere east of where we now find them.

Fig. 9-8. Outcrop with both varieties of metamorphosed sandstone in the Yellow Aster Complex, from Schermer and others, 2017. The outcrop shows a gradation between the two rock types—the contact is sedimentary, not faulted. Thus, the source area of this rock received drainage from two distinct areas.

The next step was to check out zircons in the limy sandstone, this after the 2007 paper was published. A big surprise! The zircon age pattern, measured by Liz Schermer and student Eric Hoffnagle, is completely different (right graph, Fig. 9-7); it does not correspond with a northwest Laurentian origin. Instead, the pattern matches rocks from the Appalachian mountain belt, which runs up the east coast and spreads across the east flank of Greenland. Yikes, here we are with two disparate sources!

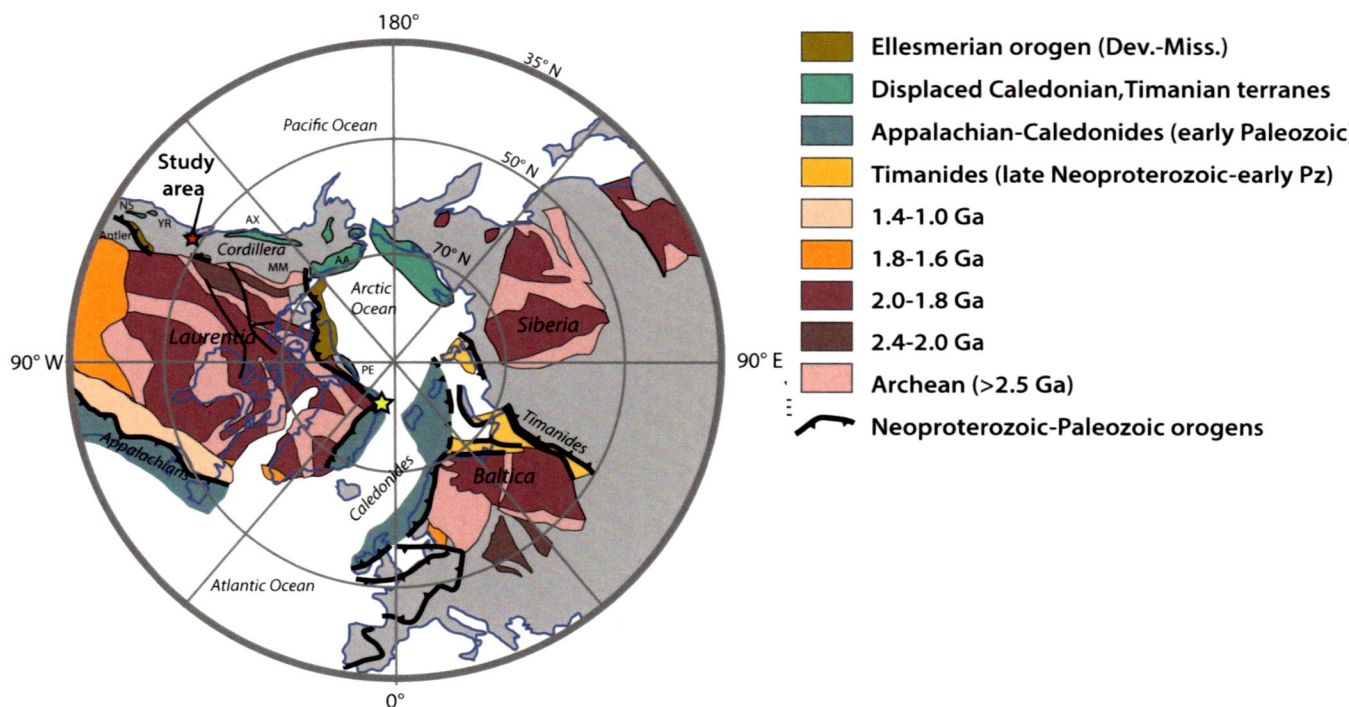

Fig. 9-9. Geologic map of craton geology in early Devonian time (~400 Ma). Colored areas are Devonian and older orogenic rocks. Grey areas mark post Devonian rocks. "Study area" is the North Cascades. The gold star marks what is thought to be the site of origin of the Yellow Aster Complex on the northeast flank of Greenland, from Schermer, et al. 2017.

Could these sandstones of different provenance have been sliced together along faults? Outcrop studies indicate a sedimentary depositional contact (Fig. 9-8). Liz Schermer picked a likely spot of origin where drainages from these two provenances could meet, in northeast Greenland (Fig. 9-9).

There are more clues in these rocks as to their origin. A well-weathered and relatively flat landscape is indicated by the very quartz-rich sands. This is not like a plate tectonic "convergent margin", but more like a "passive margin". Another finding from the sand is that the zircons have an isotope signature (hafnium ratios, that's another story) pointing to derivation from continental crust 3-4 billion years old—the very ancient parts of ancestral North America and attached eastern Greenland. But there is a further geologic happening here at about 400 Ma: development of an arc over the ancient continental margin, indicated by the igneous intrusions (e.g. Fig. 9-1). This marks the onset of plate convergence, subduction, and arc magmatism in that region.

Fig. 9-10. Model for travel of the Yellow Aster arc. The arc is initiated by subduction at the edge of the continent. The subduction zone steps back, and the arc follows its source of magma. Basalt upwelling behind the arc facilitates arc motion away from the continent and creates an oceanic backarc basin. Travel: 4 inches/year for 50 million years = ~ 3000 miles.

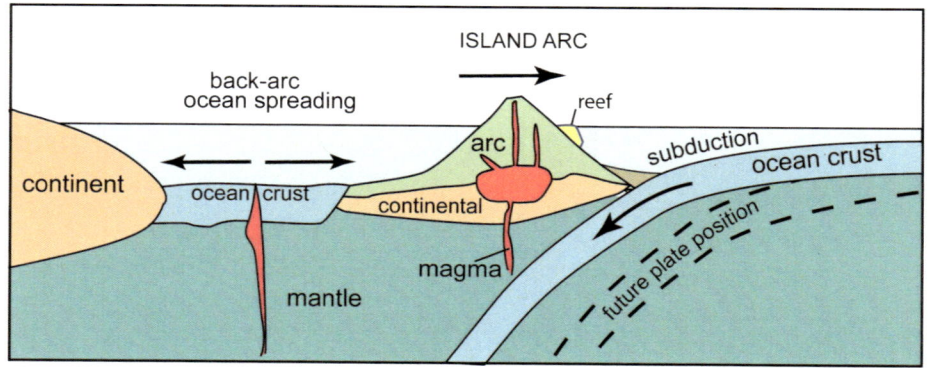

In summary, the Yellow Aster Complex brings to our doorstep a piece of east Greenland involved in a transition from a mature low-relief sedimentary plain, developed over very ancient crust, to the initiation of a continental arc at about 400 Ma.

How did this piece of a foreign mountain chain get to the Cascades? The Yellow Aster Complex is the remnant of an arc, initially built on continental crust. Most likely the arc formed at the edge of Greenland. How did it travel to western Washington? We know that ocean arcs move. The Yellow Aster arc must have been pulled off the edge of Greenland, into the oceanic realm. The dynamics of arc travels are illustrated in Fig. 9-10. The subducting oceanic plate that feeds the arc magma system is heavy enough that it tends to step back, as part of the process of sinking into the mantle. The step back pulls the magma source back, and thus the arc must follow. At the same time, mantle rock wells up in the gap left behind the arc, making basaltic ocean crust. A "back arc" sea is formed. By this means the arc moves along across ocean basins—a few inches per year. For the Yellow Aster arc, we must imagine that the initial subduction occurred at the edge of the Greenland continental crust, and that some part of the continent was captured and torn away as the arc moved seaward. Aside from understanding this mechanism in a general sense, details of the passage some 3000 miles through the Arctic and along the coast of Laurentia are beyond our sleuthing abilities at this time.

The region of the Yellow Aster Meadows and the Twin Lakes area, and beyond, offers some of the most accessible alpine meadows and peaks in the Cascades. Trails off the Twin Lakes road take you into the high country. The Twin Lakes road is rough in the switch-back area near the end, where a good clearance vehicle is recommended.

Darrington Phyllite on Goat Mountain, across the valley from the Yellow Aster Meadows (Fig. 9-4). This rock, part of the Easton Metamorphic Suite (Chapter 10), is structurally interleaved with the Yellow Aster Complex. It is mostly dark metamorphosed mudstone, rich in carbon. Peter Leiggi, barely visible, is standing on a light-colored layer of magnesium-rich carbonate rock (magnesite) ——possibly derived from mantle fragments slivered into the mudstone sediment on the ocean floor.

EASTON METAMORPHIC SUITE

Mount Shuksan seen from Picture Lake is a view broadly famous in the world. I was stunned to see this image as a wall mural in a coffee house in Hokkaido, Japan, while on a field trip there looking at rocks of very similar origin.

Key to the history of the Mt. Shuksan rocks is an exposure along the famous climbing route that passes through "Winnies Slide" —marked on photo by white arrow (Fig. 10-1). Rock there is green, indicating burial and metamorphism. Lensed blobs, separated by a dark selvedge, are squashed relics of lava pillow structure formed during eruption under water (Fig. 10-2). Chemistry of this rock indicates an origin on the sea floor erupted at a mid-ocean ridge (Fig. 1-2).

Fig. 10 -1. *above:* **Mt. Shuksan viewed from Picture Lake. The mountain is made up of metamorphosed oceanic basalt. A regional thrust fault at the base of the mountain places the Shuksan rocks over the unrelated Bell Pass Melange.**

Fig. 10 -2. *left:* **Metamorphosed and flattened ocean floor pillow basalt, now greenschist, at Winnies Slide (white arrow on mountain photo). The mineralogy of this rock indicates metamorphism in a subduction zone at depths of 15 –20 miles.**

The Easton Metamorphic Suite, of which Mt. Shuksan is part, tells us much about the mountain-building process in the Pacific Northwest. We can read in the rocks the tectonic environment and the age of formation in its original pristine condition. The subsequent tectonic processes of burial, metamorphism and uplift during plate tectonic collision are recorded in the rock mineralogy and fabric—all to be revealed by the diligent geologist in field and laboratory studies.

Coming out of Berkeley in 1966 with a PhD thesis on greenschists in Otago, New Zealand, the Shuksan belt of northwest Washington, in the backyard of my new employer WWU, was my first interest in the Cascades. The rocks are similar, but also different; comparisons could advance the science. But, the landscape is staggeringly different. Much of a field day in the North Cascades involves bushwhacking and mountaineering (Fig. 10-3). Fortunately, I liked it. I enlisted many undergraduate student assistants and graduate student researchers to help, and they liked it also.

Fig. 10 -3. View north from Helen Butte through Shuksan greenschist country. Rocky peaks to the left are parts of Mt. Watson; to the right, the glaciated peak in the distance is Mt. Shuksan; the far right mountain is Bacon Peak. All are located on Fig. 10-4. Arduous travel here; we thankfully had some helicopter support.

Regional Low-Grade Schists

Rocks of the Easton Suite occur in a variety of types, reflecting different pre-metamorphic rocks and different metamorphic and tectonic histories. The geologic map of Fig. 10-4 shows the distribution of rocks of igneous vs. sedimentary origin, and a local more strongly metamorphosed area, near Gee Point, set in a much broader field of low-grade rocks. We start with consideration of the low-grade schists. The metamorphosed igneous and sedimentary rocks occur in wide belts across the landscape, and in places on a smaller scale are interlayered. The igneous parent rock was mostly basalt, and the sedimentary parent, mostly mud and sand.

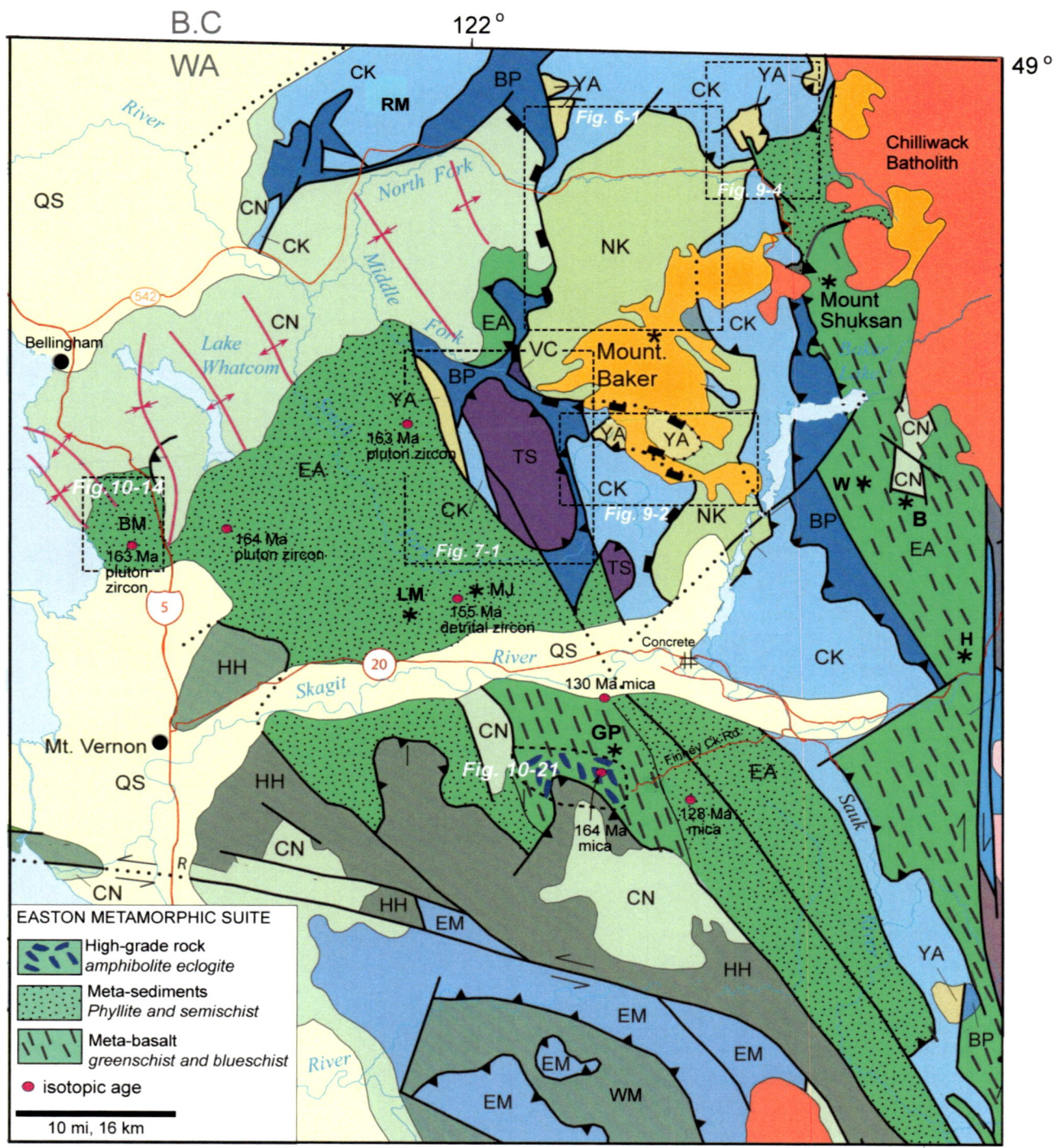

Fig. 10 - 4. Map of the North Cascades thrust system in Washington, with outlines of detailed maps that are presented in various chapters. Abbreviations of geologic units: BP = Bell Pass Melange, CK = Chilliwack Group, CN = Chuckanut Fm, EM = Eastern Melange, HH = Helena-Haystack Melange, QS = Quaternary sediment, WM = Western Melange, YA = Yellow Aster Complex. Localities of interest: B = Bacon Peak, BM = Blanchard Mtn, GP = Gee Point, H = Helena Butte, LM = Lyman Mountain, MJ = Mt. Josephine, RM = Red Mountain, W = Mt. Watson.

Contacts between these originally igneous and sedimentary rocks are largely faults. But on the south side of the Skagit Valley, a primary contact is observed. Though folded, the map relations indicate an original placement of the sediments over the basalt.

Regional metamorphosed basalt

Let's look more closely at the metamorphosed basalt of the Easton Suite. Mostly, the original sea-floor pillow structure is gone—wiped out by high strain that sheared and flattened the rock. But locally, pillows are preserved, as on the slopes of Mt. Shuksan (Fig. 10-2). Another great locality is on the side of Iron Mountain south of the Skagit Valley. The rock is totally recrystallized into blueschist, but the pillows are still discernable (10-5). Comparison with fresh unaltered pillows, as at Oamaru, New Zealand (Fig. 10-6), helps visualize the blueschist as originally a pillowed igneous rock, with a noticeable rind formed by quenching of the magma as it entered the sea.

Now, we need to consider how an originally grey-black basalt can become blue or green. The answer, of course, is the change in rock minerals from feldspar and pyroxene in the igneous rock to the metamorphic green and blue minerals (Figs. 10-7, 10-8). This is a chemical reaction driven by change in pressure and temperature. The basalt formed at ~1000 C (1800 F) and at near atmospheric pressure. The greenschist or blueschist crystallized at 400 C (700 F) and 6000-8000 bars (6000-8000 atm.)—a tremendous drop in temperature and increase in pressure. These are conditions abnormal to simple burial in the earth, and point to subduction (more on this later).

So far we have considered eruption of the Shuksan basalts in an oceanic ridge system, and subsequent subduction to form blueschists. Interestingly, there is mineralogic evidence for an intermediate step. On Mt. Watson, Ralph Haugerud during, his M.S. thesis research, discovered

Fig. 10 -5. Blueschist from the Easton Metamorphic Suite. Relict pillow structure is preserved in curved rinds below the Swiss army knife — a rare occurrence where metamorphic foliation is absent and thus has not obliterated the primary igneous structure.

Fig. 10 -6. Unmetamorphosed pillow basalt in Oamaru, New Zealand (Wikipedia), for comparison.

73

Fig. 10 -7. Microscope view of greenschist in the Easton Suite. Green minerals are mostly amphibole (variety actinolite) in this thin section. Some chlorite and epidote are also present. The black mineral is iron sulfide (pyrite), and the colorless minerals are feldspar (albite) and quartz.

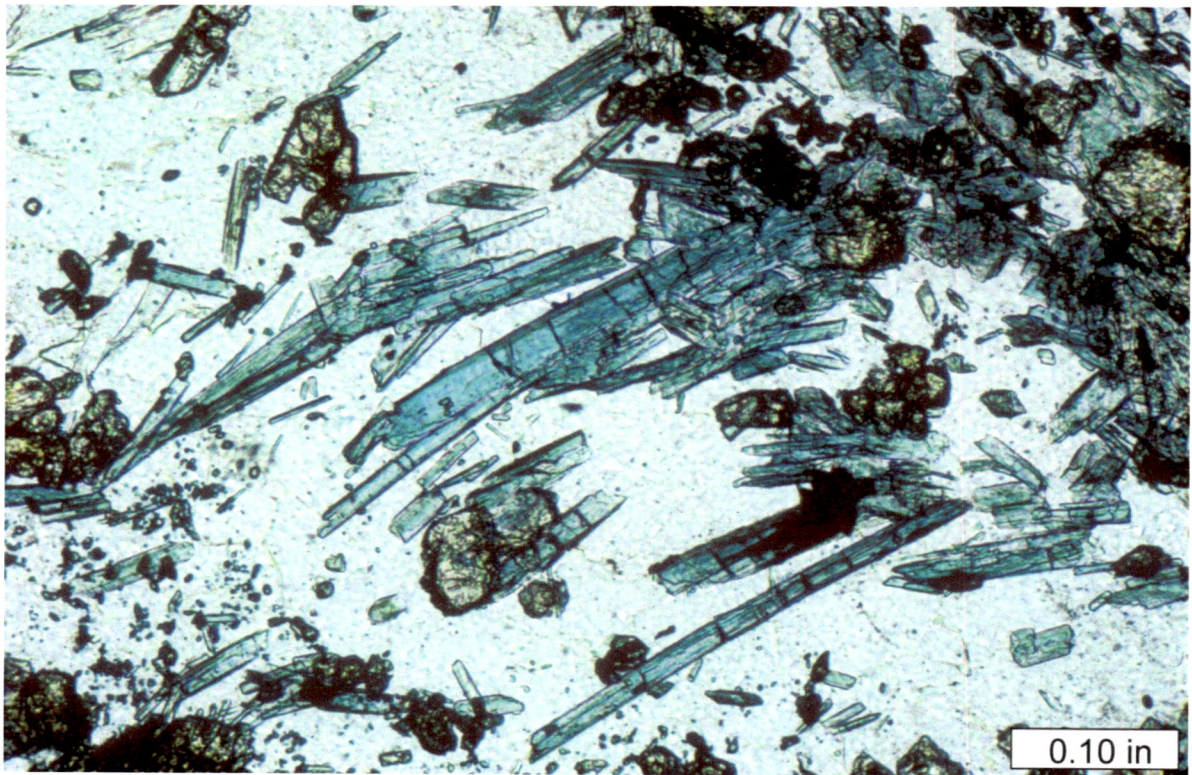

Fig. 10 -8. Microscope view of blueschist. Blue mineral is sodic amphibole (crossite). The yellow-green mineral is epidote, opaque grains are magnetite, colorless area is quartz.

globs of epidote. The epidote masses are surrounded by blueschist with strong foliation, and are pulled apart with quartz veins in the pull-apart zones (Fig. 10-9 A,B). Crossite occurs in the blueschist and in the pull-apart zones, where it is lined up parallel to the foliation in the regional blueschist. A microscope view of the epidote mass reveals a relict texture of randomly oriented mineral laths (Fig. 10-9C). This texture is typical of ocean basalts where the mineral laths are feldspar. We interpret a story for the history of this rock. It starts with hot water solutions in the active ocean ridge area that infiltrated young hot basalt. The solutions carried dissolved epidote that precipitated over and replaced the original feldspar fabric, preserving the fabric but changing the mineral. From then onward, the oceanic basalt plate was carried hundreds or thousands of miles to the ocean margin and subducted. In the subduction zone, at depth under high shear strain and great confining pressure, the basalt recrystallized everywhere to blueschist, except in the epidote blobs. The epidote blobs were pulled apart in the subduction zone, and the cross-cutting fractures filled with quartz and crossite (Na-amphibole), the crossite aligned with the regional shear fabric.

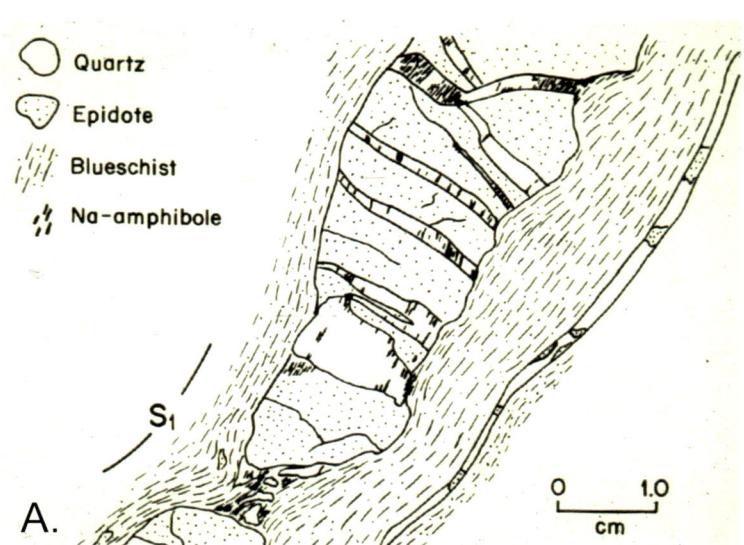

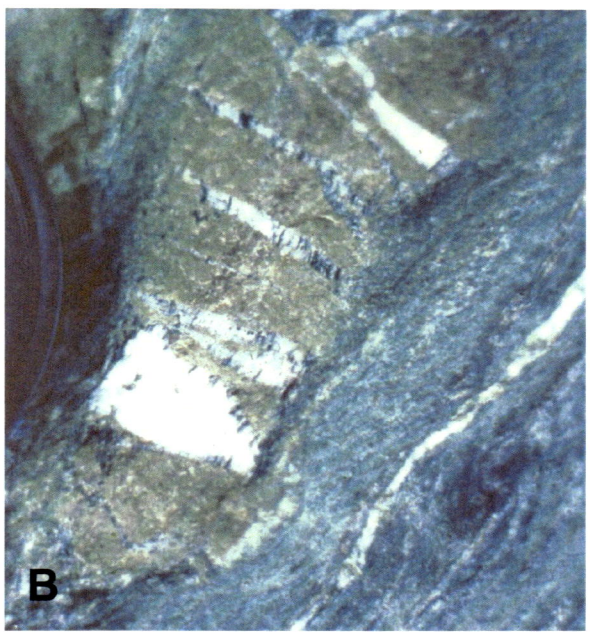

Fig. 10-9 An epidote mass in Shuksan blueschist records pre-subduction metamorphism. A and B illustrate an elongate blob of epidote in well-foliated blueschist. The epidote blob is pulled apart, making space for veins of quartz and blue amphibole (crossite) aligned parallel to the pull-apart direction and developed during subduction. C is a photomicrograph of the epidote mass revealing a ghost texture of randomly oriented mineral laths that were once feldspar in basalt, now replaced by epidote. The epidote predates high P metamorphism; it is inferred to have crystallized at the ocean ridge system. Ralph Haugerud illustrations.

Regional metamorphosed sedimentary rocks

Vast tracts of the Easton Metamorphic Suite are metamorphosed sedimentary rocks; nearly all are clastic—that is, derived from mud, sand, silt, and gravel. The total thickness is big, probably miles but not accurately measurable owing to faults that have disrupted the original sedimentary layering. These rocks conformably overlie the Shuksan meta-igneous rocks, and they show the same degree of high pressure - low temperature metamorphism.

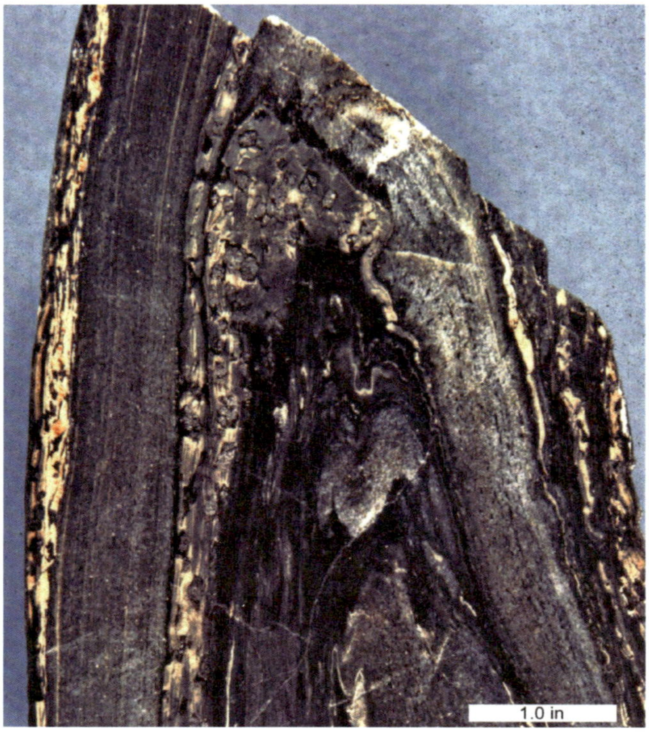

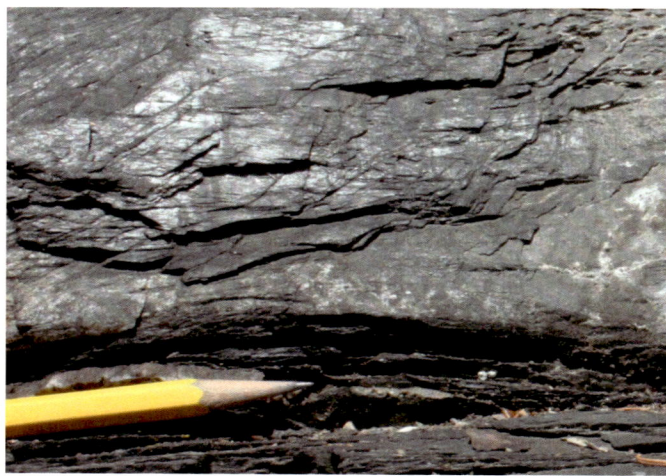

Fig. 10-10 Manganese and iron rich metamorphosed sediment in contact with underlying meta-basalt.

What can we learn about the environment in which these sediments accumulated? At the depositional contact over the Shuksan meta-basalts, the sediments are a bit weird (Fig.10-10). They are rich in silica, manganese, and iron and have crystallized tiny cream-colored manganese garnets, dark blue Fe-rich amphibole, and iron oxide. Prospectors for iron-ore long ago dug pits in the bedrock along the south side of the Skagit Valley. This chemistry apparently reflects deep sea, hot fluid emissions from the same ocean ridge volcanic vents out of which the basalt erupted.

Slightly higher in the sedimentary stratigraphy (some feet), the rock is phyllite—very fine-grained, with a prominent shiny cleavage reflecting light off fine-grained metamorphic muscovite (Fig. 10-11). Color is typically dark grey to black. The parent of this rock is mudstone.

Fig. 10-11. Phyllite. Strong metamorphic foliation formed by fine-grained muscovite crystals. Note the characteristic sheen.

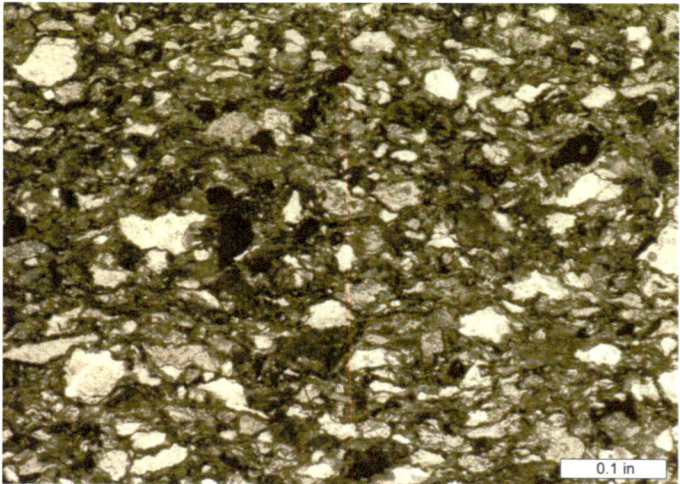

Fig. 10-12. Microscope view of metamorphosed sandstone. Colorless grains are chert and quartz; green grains are metamorphosed feldspar, pyroxene, hornblende, and volcanic rock. Zircon crystals separated for age-dating (Fig. 10-16) comprise a tiny fraction of the rock (<1%).

Belts of metamorphosed sandstone are also part of the sedimentary section (Fig. 10-12), especially plentiful in the western parts of the Easton Metamorphic Suite. Discernible grain lithologies vary regionally. Most common are feldspar, chert, and volcanic rock. We can imagine an eroded highland of mainly volcanic rock, and some uplifted deep sea-floor chert. These rocks are termed semi-schists because they are not fully recrystallized.

Relatively rare, but significant, occurrences in the sandstone region of clastic rocks are beds of metamorphosed sediments made up of extremely stretched-out pebbles, cobbles, and boulders of igneous rock (Fig. 10-13). The clasts, once equant in shape, were greatly flattened by metamorphic stress. These inherited chunks were eroded from some igneous highland, and broaden our vision of the tectonic setting of the Easton Suite, suggesting a nearby arc.

Fig. 10-13 Metamorphosed conglomerate in the western Easton Suite near Lyman Mountain (Fig. 10-4). This is a highly strained rock. White fragments are flattened clasts ranging from boulder to sand size. The white rock type is dacite, a volcanic rock of quartz and feldspar.

In looking for evidence of an arc, another rock type of the Easton Suite presents itself: masses of gabbro and granitic rock with a chemical signature of arc origin. These are scattered bodies, mostly only 10s of feet across. But, on Blanchard Mountain ("Bat Caves" locality) in the south Chuckanut region, there is an outcropping of gabbro covering several square miles (Figs. 10-14, 10-15). Is this body actually part of the Easton Suite, or is it faulted in?

Careful structural studies find that the gabbro-diorite bodies bear the same metamorphic imprint of lineations and foliation as the flanking semi-schist and phyllite. So, the gabbro could be the root of an arc intruded into the meta-sediments before metamorphism. But the absence of any intrusive features around the gabbro—dikes into the country rock and absence of contact (thermal) metamorphism of the meta-sediments—is not consistent with this idea .

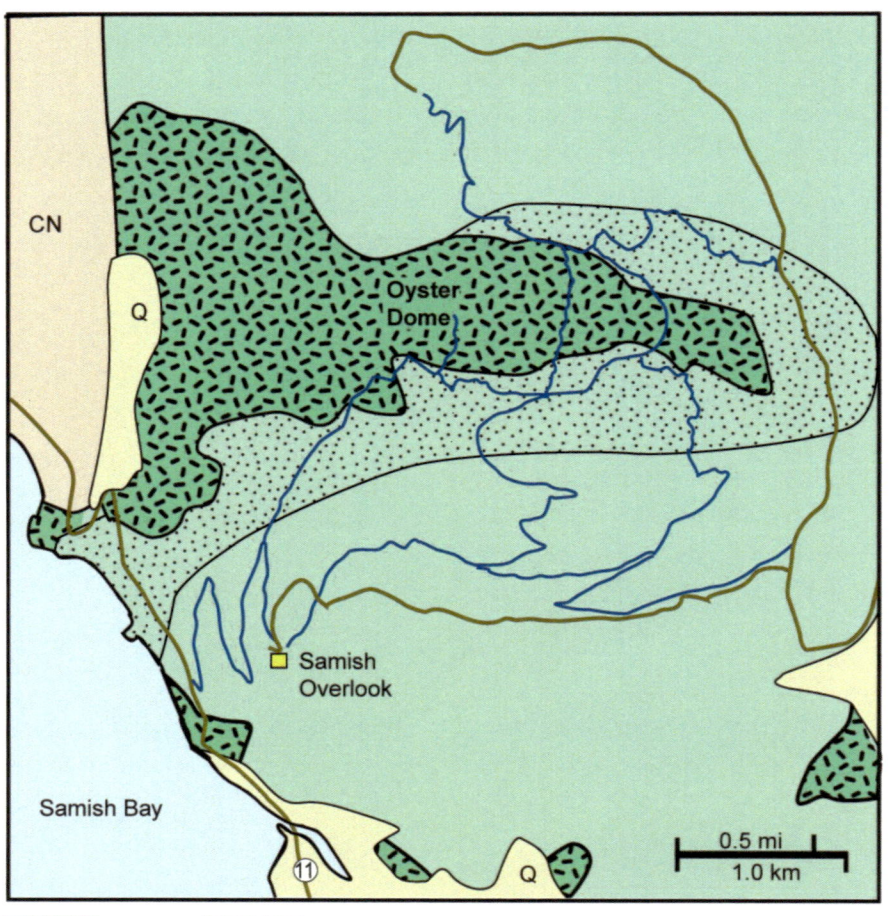

Fig. 10-14. *left:* Geologic map of the Easton Suite in the region of Blanchard Mountain (regional locality marked on Fig. 10-4). The dominant bedrock is phyllite. Gabbro occurs in numerous masses. Is the gabbro intrusive? Semi-schist, which is metamorphosed sandstone, rims much of the large gabbro mass, suggesting erosion of the gabbro as a source of the sand.

Legend:
- Phyllite
- Semischist
- Meta-gabbro
- road
- trail

Fig. 10-15. *below:* Lidar image of the south Chuckanut region, showing remarkable topographic expression of the geology. The gabbro body on Blanchard Mountain is outlined in yellow. The Chuckanut Formation (CN) shows bedding folded into a syncline, a trough-like structure. Glacially grooved moraine lies east of Blanchard Mountain.

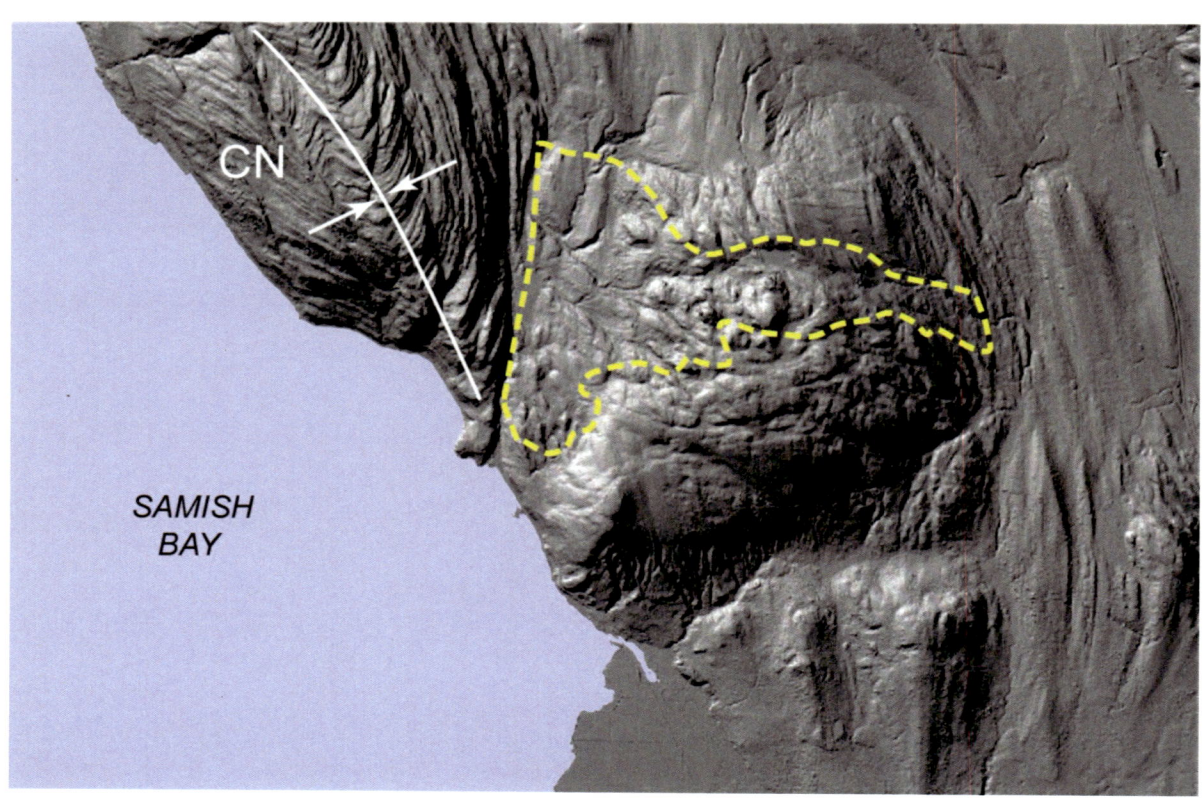

A different explanation is that this gabbro body, and other smaller igneous bodies in the region, are older than the semi-schists and were an erosional source of sand and cobbles. A spatial association of meta-sandstones with the igneous rocks, as exemplified on Blanchard Mountain, suggests this model (Fig. 10-14).

Ages of regional rocks in the Easton Metamorphic Suite

Unfortunately, any fossils that might have been in the sedimentary rock and helped with dating of the deposition are long gone due to metamorphic recrystallization. But, there is some help from zircon sand grains, which are present in very sparse concentration, but can be separated for uranium-lead dating. I collected and processed several samples of meta-sediment and came up with zircons in only one—semi-schist from Mt. Josephine (Figs. 10-12, 10-16).

This sample shows a large population at 150-155 Ma, a sparse number off the main peak in the 160-175 Ma range, and a thin population at 230-240 Ma. The maximum depositional age of this rock is about 150 Ma. Some of the zircons younger than this could be real ages, or a statistical variant of the main population.

We also know zircon ages for plutonic igneous rocks embedded in the semi-schists, indicated on Fig. 10-4. These rocks are gabbros at two localities, and tonalite at another (tonalite is a type of granite that lacks potassium feldspar). They are large chunks of rock—10s of feet to a couple of miles across, as on Blanchard Mountain (Fig. 10-14). All these plutonic rocks are associated with metamorphosed sandstone, as opposed to phyllite. And all give the same age of 163-164 Ma.

What is the history of the gabbro and granitic igneous bodies in the Easton Suite? They are apparently older than the host sandstone; thus, they can't be intrusions. We find stretched pebbles, cobbles, and boulders of granitic rock as deposits within the regional sandstone (Fig. 10-13)—and the sandstone is associated with the igneous rocks. Possibly the bodies of igneous rock arrived in the sedimentary basin as slide blocks, falling away from an old highland mass, and were eroded to produce the sand and gravel detritus of the semi-schists. However, in this process one would expect detrital zircons ~164 Ma in age, and the single sandstone sample analyzed doesn't have a distinct population of that age. A wider sampling of detrital zircon ages is needed, both to look for a 164 Ma population and also to verify that the semi-schist is in its entirety no older than 150 Ma.

Last, as far as rock ages, we have data on the timing of regional blueschist metamorphism. Potassium-argon analysis of muscovite captures the age of crystallization of this metamorphic mineral. 128 and 130 Ma are ages obtained at two sites (Fig. 10-4), marking the time of high-pressure metamorphism in a subduction zone—when the ocean crust plunged downward.

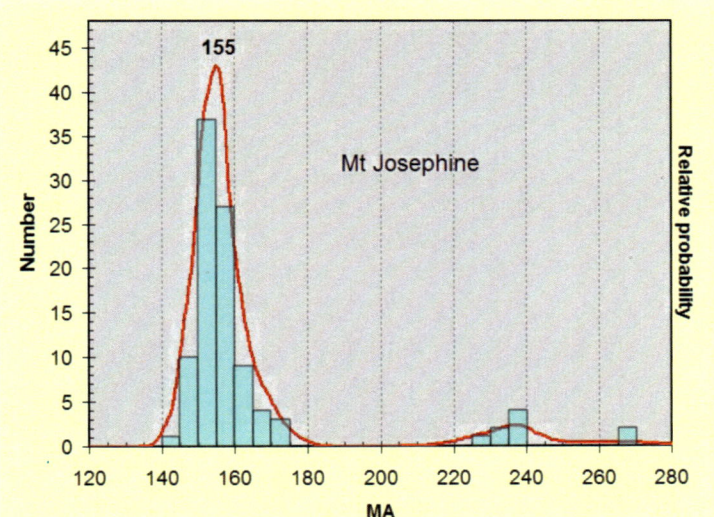

Fig. 10-16. U/Pb ages of spots on zircon grains from metamorphosed sandstone in the Easton Metamorphic Suite occurring at Mt. Josephine (locality on Fig. 10-4).

An earlier subduction event of 164 Ma is dated in the rocks of Gee Point area, discussed later.

Easton Suite metamorphic rocks bear mineralogic evidence of burial to great depths—20 miles down. The process of these rocks sliding down a subduction zone imprinted a strain fabric on the rocks, as well as new minerals (Figs. 10-17 to 10-20). At depth where temperature is elevated, and with slow strain, the rocks were plastic; they exhibit solid-state flow recorded in grain shapes and mineral alignment. You could imagine silly putty embedded with mica flakes and tiny needles. With slow flattening and smearing out on a hard surface, the putty would thin out and the micas would align in the plane of flattening. The needles would align both in the plane and also in the direction of smearing out. Rocks record this type of deformation, and the structural fabric we see tells us how the rocks moved. In turn, we can relate the fabric we see in rocks spread over a region to the plate tectonic motion that smeared them out. The smear direction and sense of motion indicated for the Easton Suite, across a wide area, point to a subducting plate moving east-northeast (Fig. 10-20). Given that the age of metamorphism is dated at 128 and 130 Ma, we can piece together a model for plate tectonics of the region.

Fig. 10-18. Smeared and aligned relict igneous feldspar crystals in metamorphosed basalt of the Easton Suite near Anderson Lake.

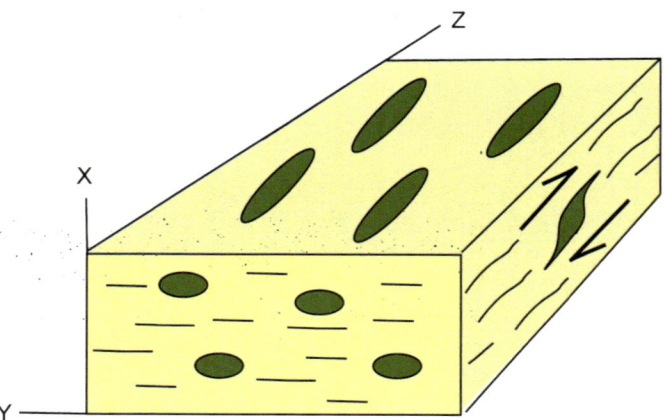

Fig. 10-17. Simple view of shear strain in a rock modelled on a deck of cards. Smear the cards out in the Z direction—this is shear. If there is a marker fragment in the shear zone (in the case of a rock this could be a sand grain) the marker is elongate in the shear direction. On the XZ plane, we would see an asymmetry that records the rotation sense of the upper cards moving over the lower ones.

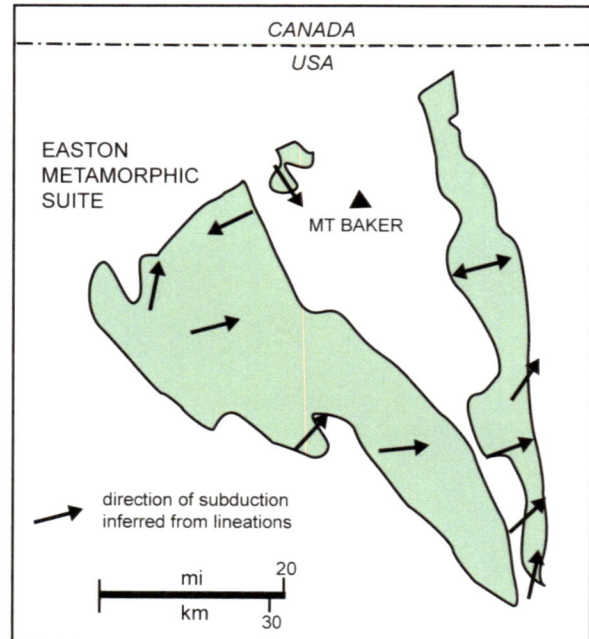

Fig. 10-20. Metamorphic lineations in the Easton Suite. Direction and shear sense indicate subduction in an east-northeast direction.

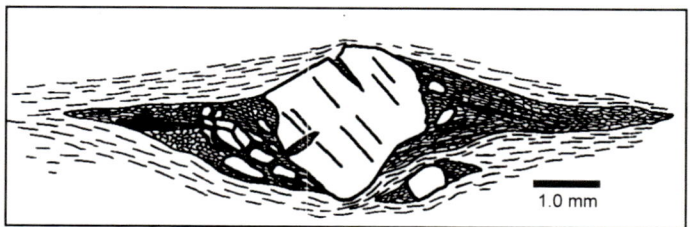

Fig. 10-19 Relict feldspar grain broken and rotated clockwise by shear. View on XZ plane. Greenschist in the Easton Suite.

Gee Point High-Grade Rocks—Initiation of Subduction

Most of the Easton Suite has mineralogic evidence of relatively low temperature metamorphism (300-400 °C), and some sedimentary and igneous textures can still be seen, giving us a window into part of the plate tectonic history of the region. Much higher grade rocks are found in the hills on the south side of the Skagit Valley in the vicinity of Gee Point (Figs. 10-4, 10-21, 10-24), revealing an earlier part of the subduction zone system. Here we find the subducted oceanic slab metamorphosed to very high temperatures and pressures—evidence of this rock diving into the earth's mantle at hotter conditions than for the low-grade rocks described above.

In mapping around in this region in the late 1970s, we found the unusual circumstance of belts of serpentinite interleaved with high-grade metamorphosed oceanic rocks (Fig. 10-21). A first explanation was the likelihood that the serpentinite was part of the Haystack Melange lying just to the south of the Easton Suite (Fig. 10-4) and was faulted into the Easton during thrust emplacement of the Haystack nappe, long after Easton metamorphism. But, on further inspection of the contact between serpentine and the meta-oceanic rocks, it was apparent that the juxtaposition was not a late, shallow fault, but that it happened at great depth in the subduction zone during high-grade metamorphism. The serpentinite is in tight contact with amphibolite formed from oceanic rocks (Fig. 10-22). High-grade mineral zones are developed between the two rock types, marking a transition from ocean crust to mantle rocks.

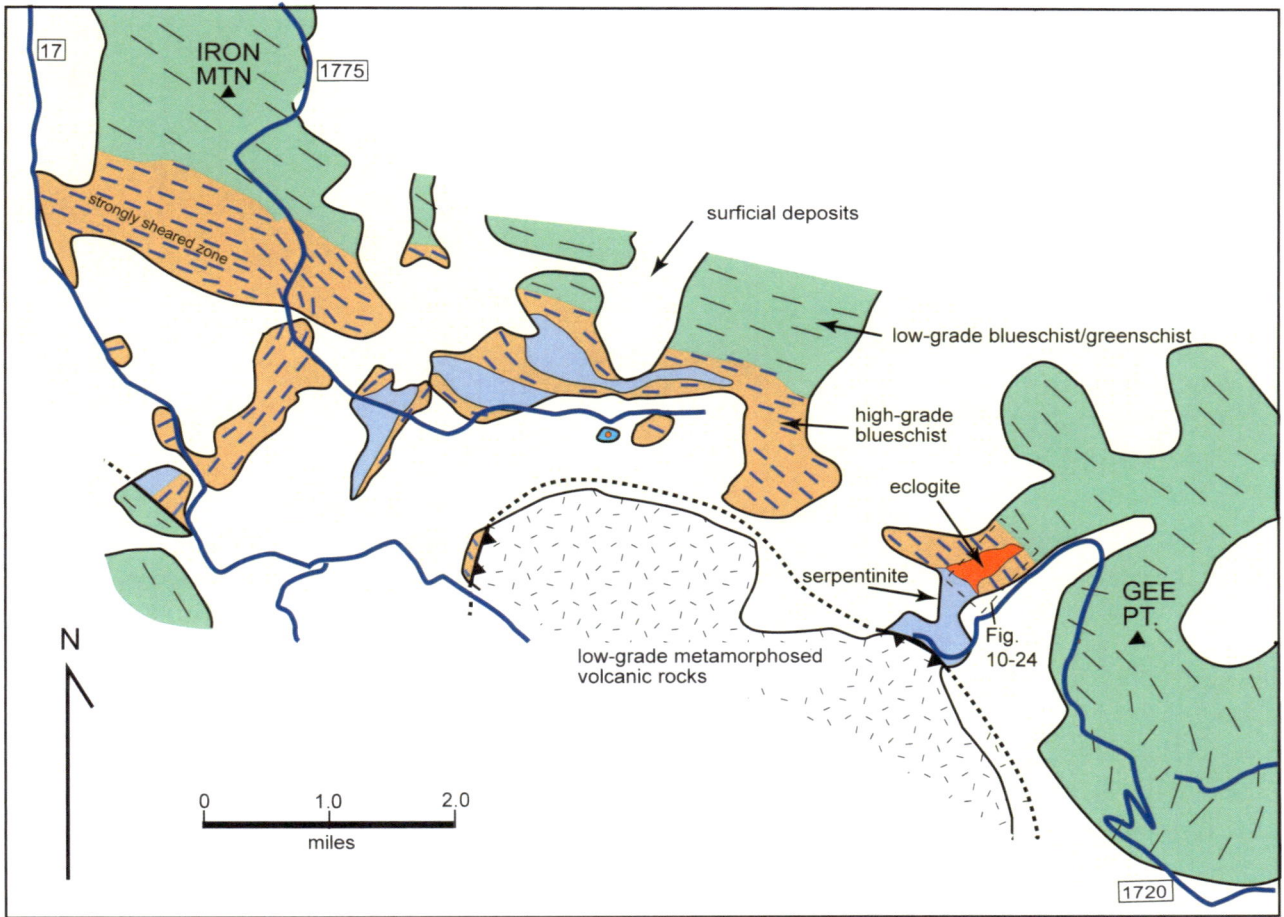

Fig. 10-21. Geologic map of the Gee Point - Iron Mountain area, lying just south of the Skagit River. See location on Fig. 10-4. In this region is an assemblage of high pressure *and* high temperature rocks inferred to represent the initial stage of subduction of the Easton Suite oceanic rocks.

The serpentine was derived from olivine—the outlines of which are still visible (Fig.10-23). Relict chromite once associated with the olivine is still preserved in the serpentinite, and its composition together with that inferred for the olivine gives a temperature of about 800 °C. Minerals in the metamorphic oceanic rocks likewise give temperatures up to that magnitude. Recrystallization of olivine to serpentine happens in the presence of water at temperatures below 600-700 °C.

The olivine rock, parent to serpentinite, is almost certainly of mantle origin. A close relative in which the olivine is still preserved is the Twin Sisters dunite featured in chapter 7.

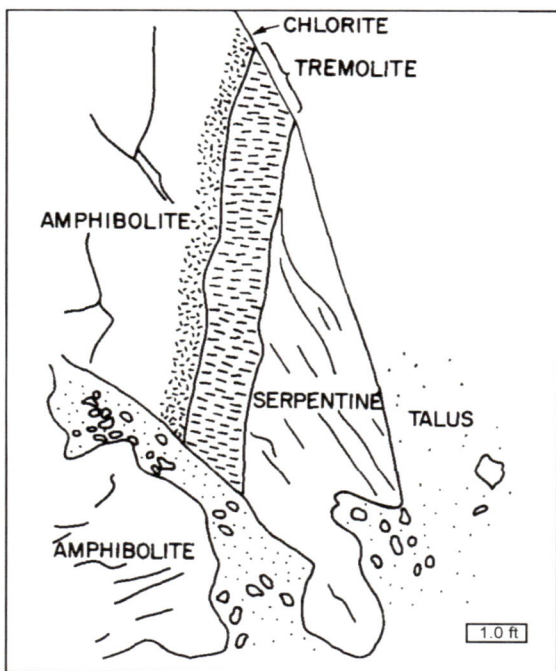

Fig. 10-22. Roadside outcrop of the contact between meta-ocean crust (amphibolite) and rock of the earth's mantle (serpentine). Chlorite and tremolite zones formed by exchange of chemicals during high-grade metamorphism.

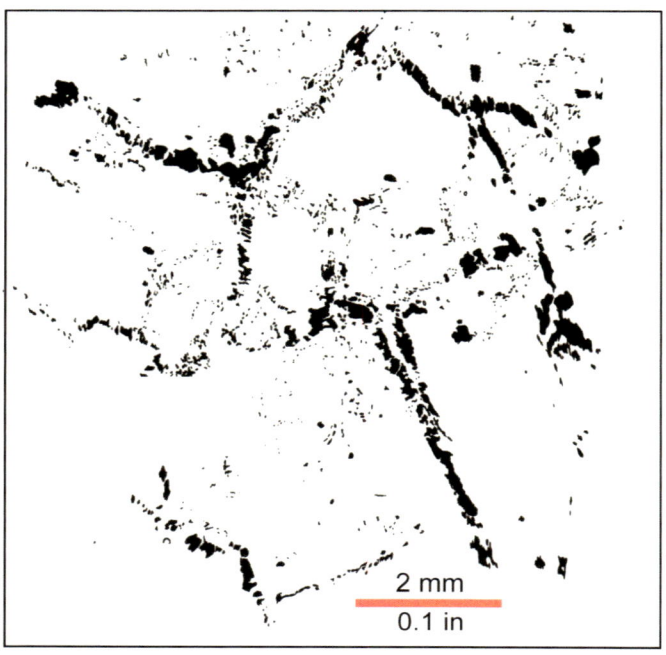

Fig. 10-23. Drawing from a microscope view of serpentinite rock. Black grains are magnetite formed during alteration of olivine to serpentine. As iron-poor serpentine replaced the olivine, iron moved out to the olivine grain boundaries and crystallized as magnetite—outlining the shape of the original olivine grains.

An exceptional exposure of the high-grade ocean floor rocks and intermixed serpentinite is along a ridge just west of Gee Point (Figs. 10-24, 10-25). This locality, exposed by logging and accessed by forest roads in the 1970s, is now unfortunately overgrown and part of the forest road is washed out.

The ocean floor rocks are relatively intact, but totally remade into metamorphic rocks. Ocean floor basalt is now coarse-grained amphibolite and eclogite. Sea floor chert is recrystallized to quartzite. Clay-rich deep sea ooze is transformed to coarse-grained muscovite schist. An original stratigraphic sequence of these rocks is preserved (Fig. 10-26). High-grade mineral assemblages include garnet, hornblende, and omphacite (Fig. 10-27, 10-28).

In addition to these metamorphic rocks that pretty much preserve the bulk chemistry of the sea-floor rocks, there are other metamorphic rocks that have weird mineral assemblages and textures (Fig. 10-29). These rocks are what we call metasomatic. They crystallized from water-rich fluids moving to and from the metamorphosed ocean-floor and serpentinite hot rock masses.

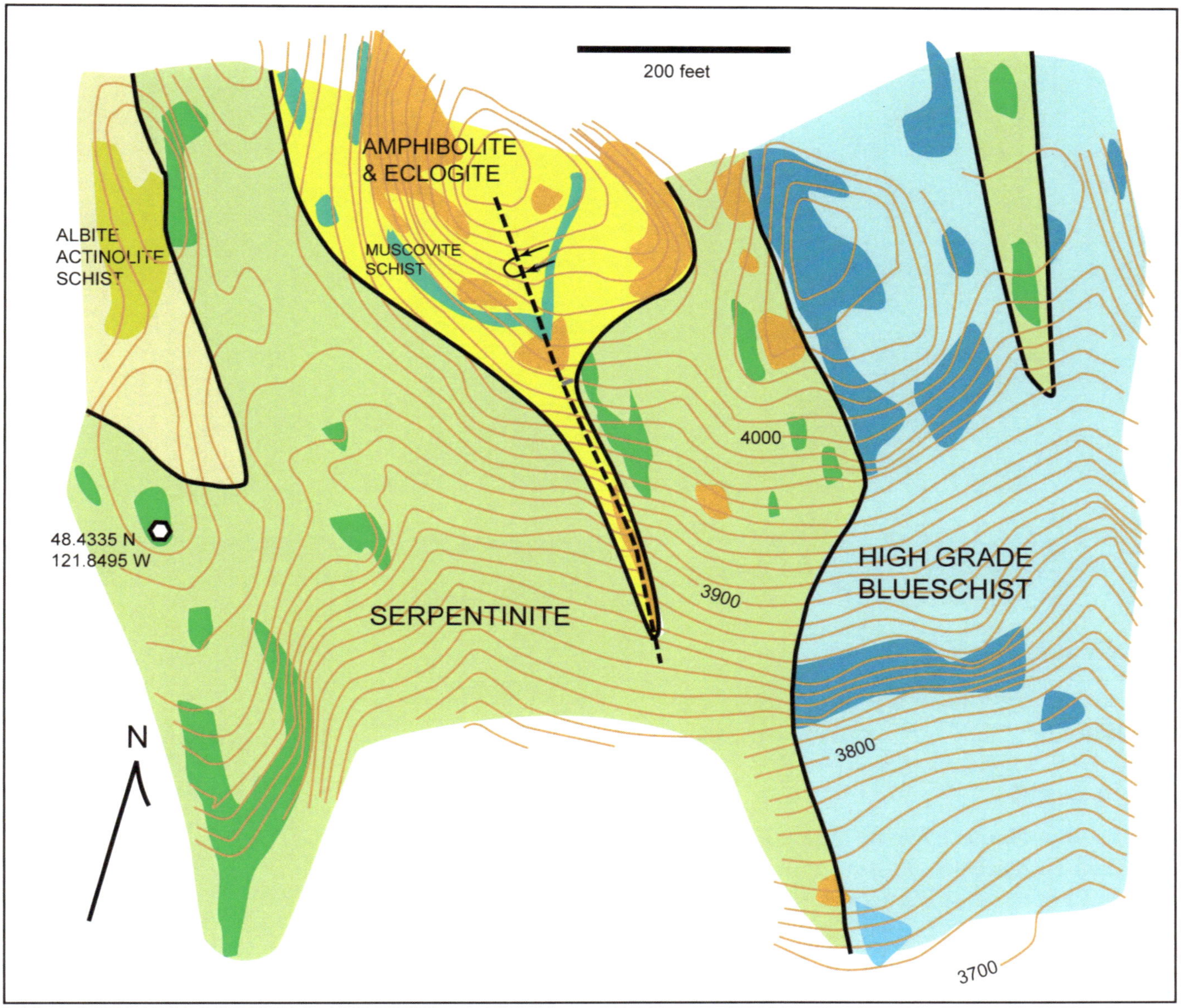

200 feet

AMPHIBOLITE
& ECLOGITE

ALBITE
ACTINOLITE
SCHIST

MUSCOVITE
SCHIST

4000

48.4335 N
121.8495 W

HIGH GRADE
BLUESCHIST

SERPENTINITE

3900

N

3800

3700

Fig. 10-24. *above:* **Map of Gee Point locality where exceptional field relations of metamorphosed ocean floor rocks (blueschist, amphibolite, eclogite) and mantle derived peridotite (now serpentinite) are preserved.**

Fig. 10-25. *left:* **View of the clear-cut ridge in the map above, 1980. At that time there was good exposure and road access. Now the locality is overgrown, and the road washed out.**

What is the relationship of the Gee Point high-grade rocks to the regional, lower grade blueschist and greenschist? One critical evidence comes from the radiometric ages. K-Ar dating of mica in the lower grade rocks yields a 128-130 Ma age. Dating of the high-grade mica in the Gee Point area yields an age of 164 Ma, a significantly older age, well beyond analytical uncertainties.

Fig. 10-26 *above:* **Preserved ocean-floor sequence of (originally) basalt overlain by chert, in turn overlain by clay-rich ooze. Now: amphibolite, quartzite, mica schist.**

Fig. 10-27. *right:* **Garnet-hornblende amphibolite. High-grade metamorphic rock. All vestiges of original sea floor basalt are gone.**

Fig. 10-28. *below:* **Crossite (blue)-omphacite (green) rock. Both minerals indicate high-pressure metamorphism. The omphacite is a Na-bearing pyroxene, distinctive of the rock eclogite.**

Textural information backs up the age difference. In many places, the high-grade rocks show invasion and replacement by lower grade minerals, and an overprinting by younger foliation (Fig. 10-30) that is developed on a regional scale in the low-grade rocks.

So, what kind of a plate tectonic explanation can we come up with for the Gee Point rocks and their relationship to the lower grade blueschist/greenschist and phyllite/semi-schist of regional extent? We know the

Fig. 10 -29. A thick vein of very garnet-rich rock is developed in high-grade garnet blueschist. Hot, mineral-rich watery fluid deposited this vein, replacing the blueschist. A technical term for this process is metasomatism.

following useful facts: 1) Metamorphism of the Gee Point rocks is older —164 vs. 130 Ma. The protolith of the regional rocks, from one sample, is younger, < 150 Ma from detrital zircon (Fig. 10-16), than the 164 Ma metamorphism in the Gee Point rocks. 2) Rocks in both groups are of sea floor origin. 3) Metamorphic pressures in both suites of rocks are similar at about 7-9 kilobars, equating to ~25-35 km (15-22 mi) burial. A subduction zone environment is envisaged for both. 4) Metamorphism occurred at notably higher temperature in the Gee Point rocks (~500 –700 °C) than in the younger regional rocks (~350-400 °C). 5) And, last but not least in the list of important facts, the Gee Point assemblage is intimately mixed in with serpentinite derived from the mantle—here is a record of ocean floor rocks driven into the mantle.

We can make a credible story out of the above observations. In the beginning was open ocean stratigraphy: ocean crust made of, from top to bottom, sediments, basalt, gabbro (Fig. 10-31, stage 1). Temperature increased downward at a typical earth rate of about 25°C per km. Then, the crust was

Fig. 10 -30. Sketch of texture in high-temperature rocks that have been partially overprinted by lower temperature assemblages. In this sample the lower-grade foliation is defined by newly grown chlorite, muscovite, and crossite. The foliation wraps around and has caused breakage of the relict garnet and hornblende. This fabric and mineralogy correlate with the regional relatively low-grade blueschist metamorphism.

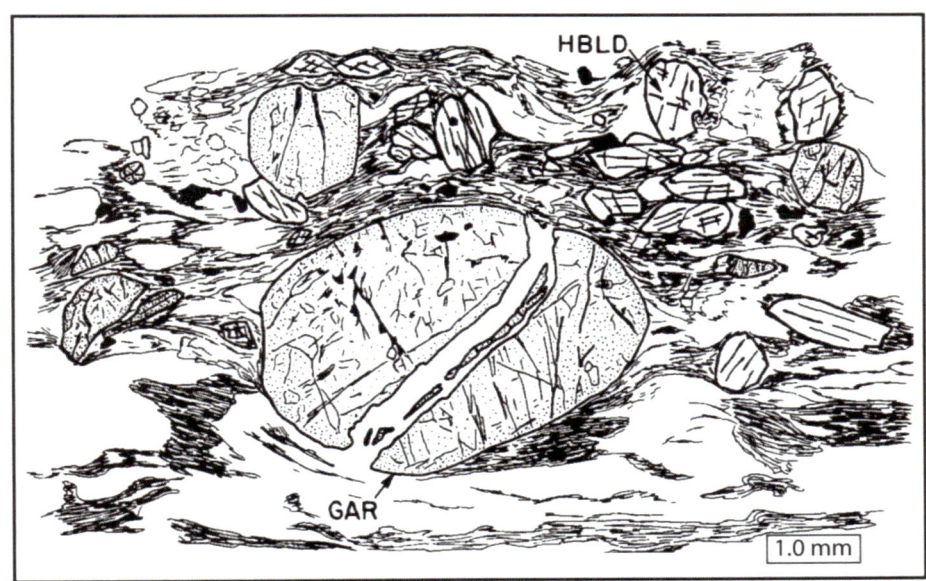

85

broken. A subduction zone was initiated (stage 2); the ocean crust plunged downward. Relatively cold surficial rock was pushed into hot rock. The mantle isotherms were bent downward, and the subducted rock was heated. Slabs of mantle serpentine (or peridotite) were sliced into the ocean rock. In this deep, hot zone the ocean rocks were totally remade into high-grade metamorphic rocks. Much water was released from lower grade rocks as the hydrous minerals they carried broke down with increasing temperature. Interesting water-deposited mineral assemblages formed metasomatic zones.

The next step (stage 3) is a failure of the initial subducting slab to penetrate more deeply. The subduction zone steps back and sends another ocean slab down, this one just under the high-grade rocks which are now accreted to the overlying plate. The newly subducted crust pushes far into the mantle. The subducted cold crust greatly deflects mantle isotherms, providing a broad zone of pressure-temperature conditions favorable to crystallization of low-grade blueschist, as we find on a regional scale in the Easton Suite. In this process, the Gee Point rocks are somewhat overprinted by the later low-grade mineralogy.

Plate Tectonics of the Easton Suite

Finally, on a crest of wild speculations, let's think about a plate tectonic setting for the Easton Suite as a whole, that broadly considers the spectrum of minerals and structures described in this chapter. Fig. 10-32 is a model. It shows an ocean ridge spreading center where pillow basalt magma erupts, crystallizes, and adds to the edge of ocean crust. Ocean water circulates into the eruption zone, dissolving and depositing iron and manganese that form sedimentary layers. Hot water also overprints basalt textures with epidote. As the ocean plate travels off the ridge center, chert is formed by the settling of radiolaria.

Subduction is toward the east, deduced from strain features formed in subducted rock of the low-grade schist (Fig. 10-20). Subduction

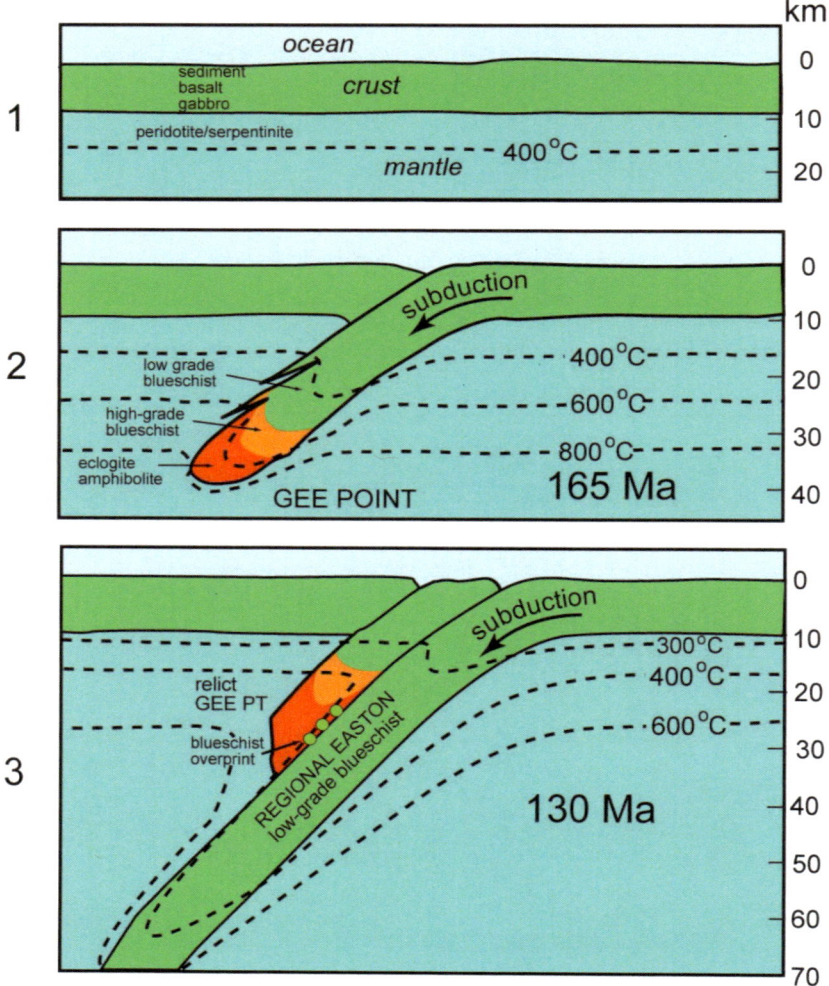

Fig. 10 -31. Hypothetical stages for the plate tectonic history of the high-grade Gee Point rocks and regional low-grade blueschist metamorphism. Stage 1 predates subduction. Stage 2 is initial subduction in which the ocean crust first runs into hot mantle rock, yielding high-grade metamorphism. In Stage 3, the subduction zone steps down, stranding the high-grade rocks, and bringing ocean crust into a realm of high pressure but lower temperature than in the initial phase of subduction.

began at roughly 165 Ma, marked by the high-grade Gee Point rocks, the first to encounter hot mantle. Subduction stepped back, but carried on through 130 Ma, creating the regional low-grade blueschist rocks under-thrust into mantle already somewhat cooled by the initial subduction.

But what about the addition of the extensive sedimentary deposits of the Darrington Phyllite? Most of this rock was deposited as mud, now phyllite, but a fair bit also was sand, and some even coarse enough to be called conglomerate. Identifiable grains include chert fragments, but most are igneous fragments, both volcanic and plutonic. The one sample analyzed for detrital zircon gives a major population of 150 -155 Ma, which could define the age of the most abundant igneous rocks in the highland. We envisage this highland, shedding sediment over the oceanic rocks, to lie to the east in an arc over the subducting slab. Much of the sediment is subducted. A kink in this model comes from widely scattered large chunks of plutonic rock 10s of feet to miles across, gabbro and granitic rock, that give igneous zircon ages of 163-164 Ma, described earlier. These rocks occur mostly in association with sandstone and conglomerate. They could be slide blocks, possibly coming off the main structure of the arc volcano that we postulate to have been active since ~165 Ma. More work is needed.

So, we are saying that the Easton Suite is the remnant of ocean crust and an oceanic arc. The extensive sedimentary section, Darrington Phyllite, formed in a basin in front of the arc. This would be termed a forearc basin. Where was this geology happening with respect to the North American continent? One test is the presence or absence of North American Precambrian zircon in sediment. For the one sample so far studied, such zircons are not found. As far as we know, behind the Easton Suite arc was ocean, and North America somewhere beyond. Where the arc developed north or south, outboard of the continent, is unknown. All we are sure about is that the rocks arrived where they are now by much dislocation.

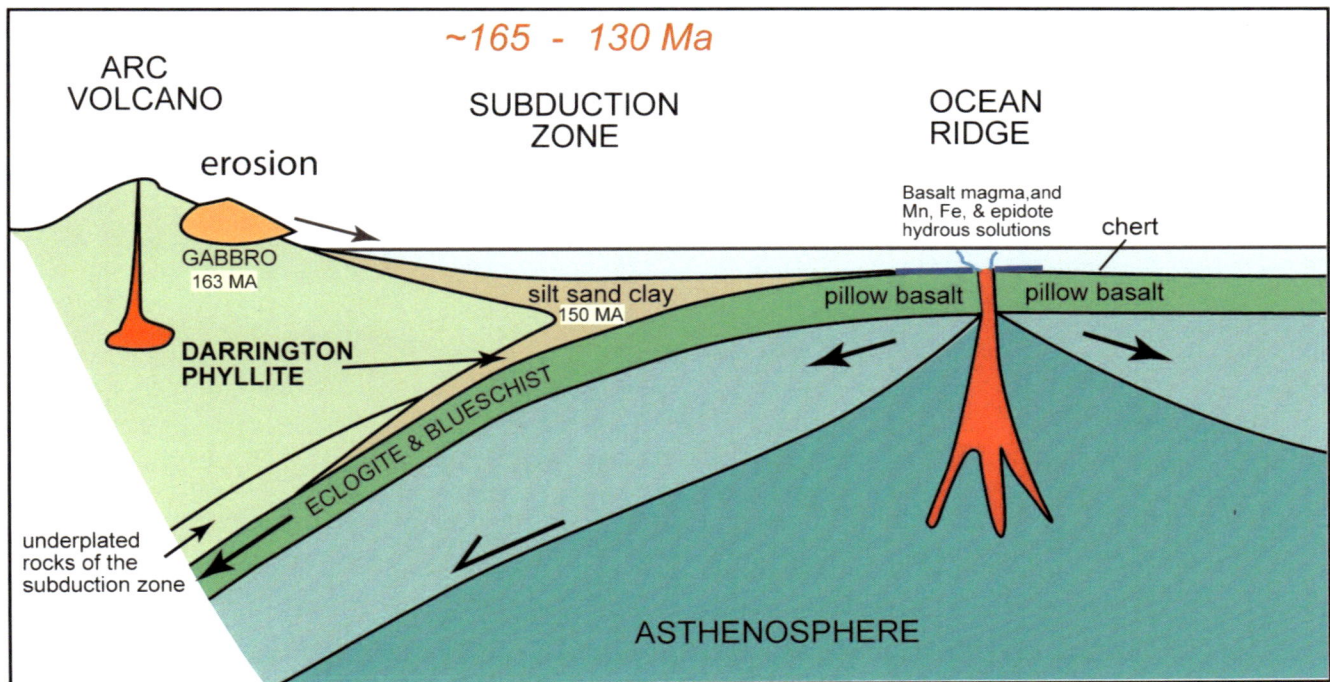

Fig. 10 -32. Regional tectonic model for the origin of the Easton Suite. View south. Did the gabbro body of the Oyster Dome (Fig. 10-14) slide into the sedimentary basin?

Part III — SOUTHERN COAST PLUTONIC COMPLEX

CHAPTER 11

HARRISON LAKE AREA

Introduction

The Coast Plutonic Complex (Figs. 4-1, 4-2, 11-1) is a belt of granitic intrusions (plutons) and country rock terranes that extends from Washington to Alaska, roughly a thousand miles. The width of the belt varies from about 50 to 100 miles. The age of pluton intrusion spans a time frame from 170-50 Ma. The total volume of these plutons is vast, ~40,000 cubic miles assuming a pluton depth of one mile (very conservative), and equal volumes of pluton and country rock. An additional approximately 30% of this volume was erupted as volcanic rock, now eroded. Where did this immense amount of magma come from, and how did plutons make space for themselves in the host country rock at our continental margin?

Fig. 11-1. View north along alpine exposures of the Urquhart pluton in the Harrison Lake area.

Our area of interest for this book includes the southern B.C. Coast Mountains in the vicinity of Harrison Lake, and the North Cascades "Crystalline Core". These two southernmost parts of the Coast Plutonic Complex were coextensive before displacement on the Straight Creek Fraser River fault. (Figs. 4-2, 4-3).

The Coast Plutonic Complex in these two regions documents episodes of crustal thickening by as much as 20 miles or more, approximately coeval with the plutonism. Good explanations of this process remain elusive. In chapters 11 and 12, we will be immersed in an interplay of plutonism, metamorphism, and plate tectonics. There's a lot to think about.

A. TECTONIC LOADING

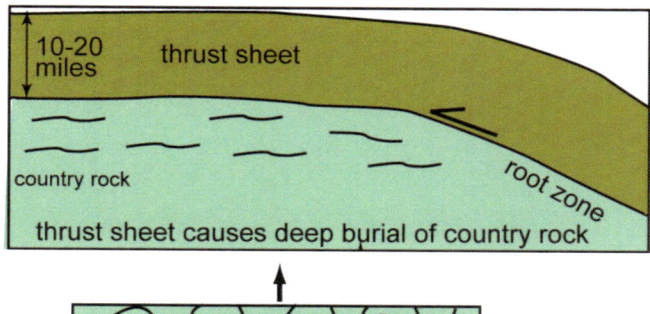

10-20 miles — thrust sheet

country rock

thrust sheet causes deep burial of country rock

root zone

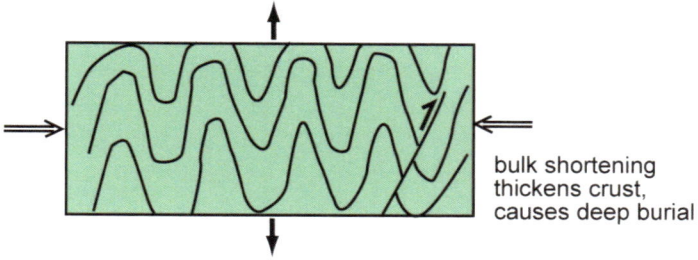

bulk shortening thickens crust, causes deep burial

B. MAGMA LOADING

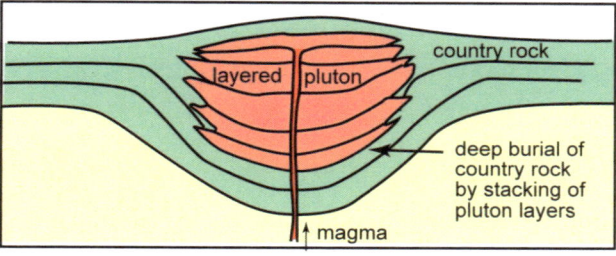

country rock

layered pluton

deep burial of country rock by stacking of pluton layers

↑ magma

Fig. 11-2 Tectonic vs. plutonic models for the mountain-building origin of the Coast Plutonic Complex.
A) Tectonic contraction occurs across the arc. In one scenario, contraction forces a thrust sheet to rise up out of the region and slide over the country rock, which is then buried by as much as 20 miles or more. The "root zone" marks a discernable spot from which the thrust originates. Another possible tectonic mechanism is contraction accommodated by "bulk shortening" in the form of relatively small tight folds and reverse faults; the crust thickens.
In both tectonic scenarios, plutons are incidental.

B) Pluton emplacement causes crustal thickening. Plutons grew by injection of horizontal sheets, accumulating to a thick stack. Magma streamed up from the base of the crust and spread out. Many successive layers thickened the crust.

There is general agreement among the dozen or so geologists focused on this problem that the Coast Plutonic Complex is a magmatic arc, formed by melt rising out of the subduction zone of the Farallon Plate under the western edge of North America. However, explanations for the great crustal thickening accompanying the arc magmatism, and consequent mountain building, are debated.

One interpretation is that crustal thickening, and burial of country rock and plutons, 10 to 20 miles, occurred by tectonic contraction (Fig.11-2A). Two possible scenarios are: 1) thrusting of a great sheet over country rock, and 2) bulk shortening of country rock, and consequent thickening, by tight folding and small-scale thrust faults. Plutons are emplaced before, during, and after tectonism, but are incidental to the crustal thickening process.

An alternative interpretation is that the plutons grew in the upper reaches of the crust by stacking of granitic layers, each of which was fed by a stream of magma rising from a reservoir at the base of the crust. In this mechanism, crustal thickening, and high-pressure metamorphism of country rock, is tied to pluton growth. Crustal contraction is not part of the process (Fig. 11-2B).

Why is this issue of how the Coast Plutonic Complex grew important? The CPC is a huge orogenic belt. It represents a fundamental process of mountain building and continental growth. Likely, what occurred in the CPC is relevant to other mountain belts.

With this thought in mind, that understanding the origin of the Coast Plutonic Complex is important, these two chapters in Part III go into some detail about solving the problem of which process prevailed, tectonic contraction or horizontal pluton sheeting (Fig. 11– 2).

Here we need to bring out our "tools" of Chapter 3 for measuring temperature, pressure, and age of formation of rocks. We also need to consider field relations, including map patterns and outcrop features. Rock lineations are important, as they give indication of which way ductile rock was translated (e.g. Fig. 10-17).

From a lot of detail outlined in this chapter and the next, my conclusions are that: 1) A good case can be made for model **B** in the Harrison Lake area. 2) For the Cascades Crystalline Core, clear evidence exists for model **B** in one pluton (Golden Horn). But for the rest of the Core, the crustal thickening process is debatable; **A** vs. **B** remains uncertain. For the non-geologist, Chapters 11 and 12 will be challenging. Hang in there.

Pluton Emplacement Mechanisms

A variety of potential pluton emplacement mechanisms is illustrated in Fig. 11-3. Controversy and uncertainty about how plutons form stems from the fact that we typically have a 2-D view of a 3-D feature, and that intrusive igneous rocks commonly lack much in the way of mineral alignments or other fabric that reveals intrusive flow patterns.

The *diapir* model is popular, but how common is uncertain. Magma floats up in a blob, like an air bubble rising through water. The country rock is plastic and flows around the rising magma chamber. This flow strain is imprinted on the country rock in a zone of some breadth that follows the perimeter of the pluton. A *dike* is a tabular body; the magma moves into a plane of weakness in the country rock. A *sill* (not shown) would be a tabular body intruded parallel to fabric in the country rock. In the *stoping* mechanism the magma blob rises by breaking off chunks from the roof that fall to the floor; in this case, the pluton is riddled with country rock inclusions. Pluton emplacement by *sheeting* yields a layered mass. If the sheets are vertical, then the country rock has accommodated the pluton by spreading laterally. Where the sheets are horizontal, the country rock has spread up or down. Finally, we come to *ballooning*. Magma rises through a fissure to a point where a growth zone is initiated. The pluton grows in place by magma inflow. Country rock is displaced: out in all directions if the growth zone is a blob, sideways if the growth zone is a vertical dike, or up/down if the growth zone is a horizontal sill.

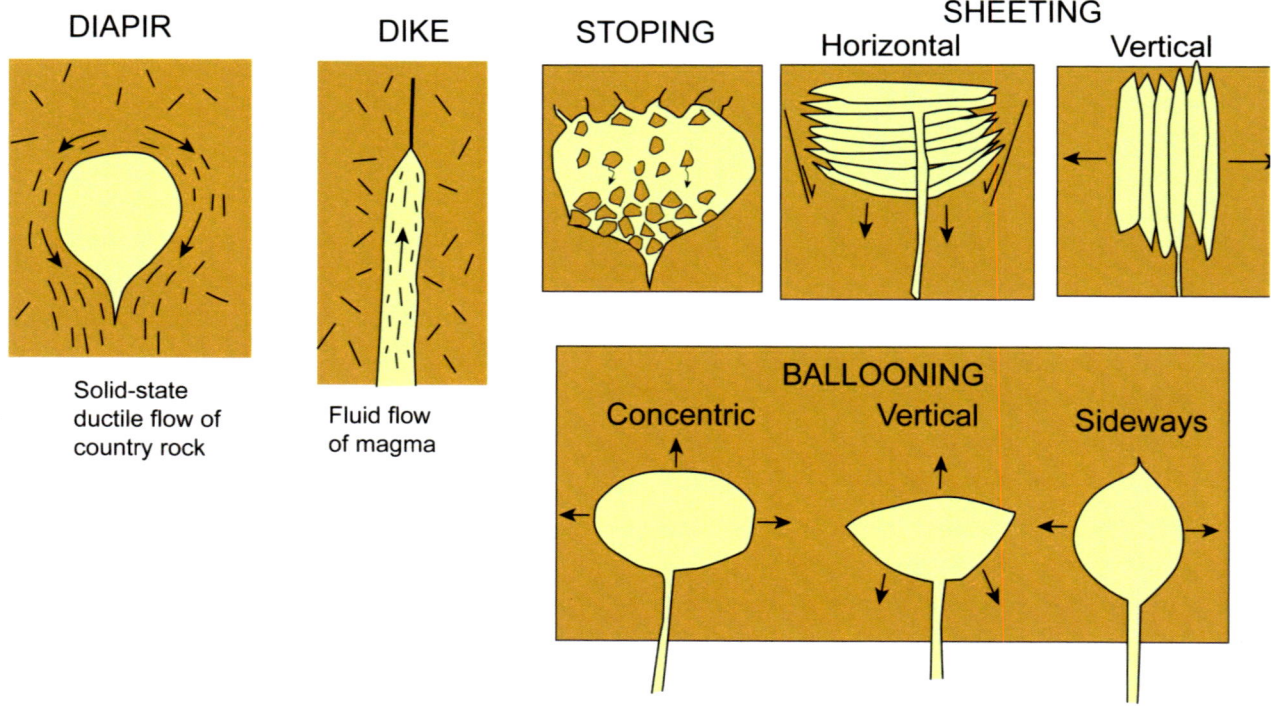

Fig. 11-3 Cartoons illustrating emplacement mechanisms for plutons invading country rock.

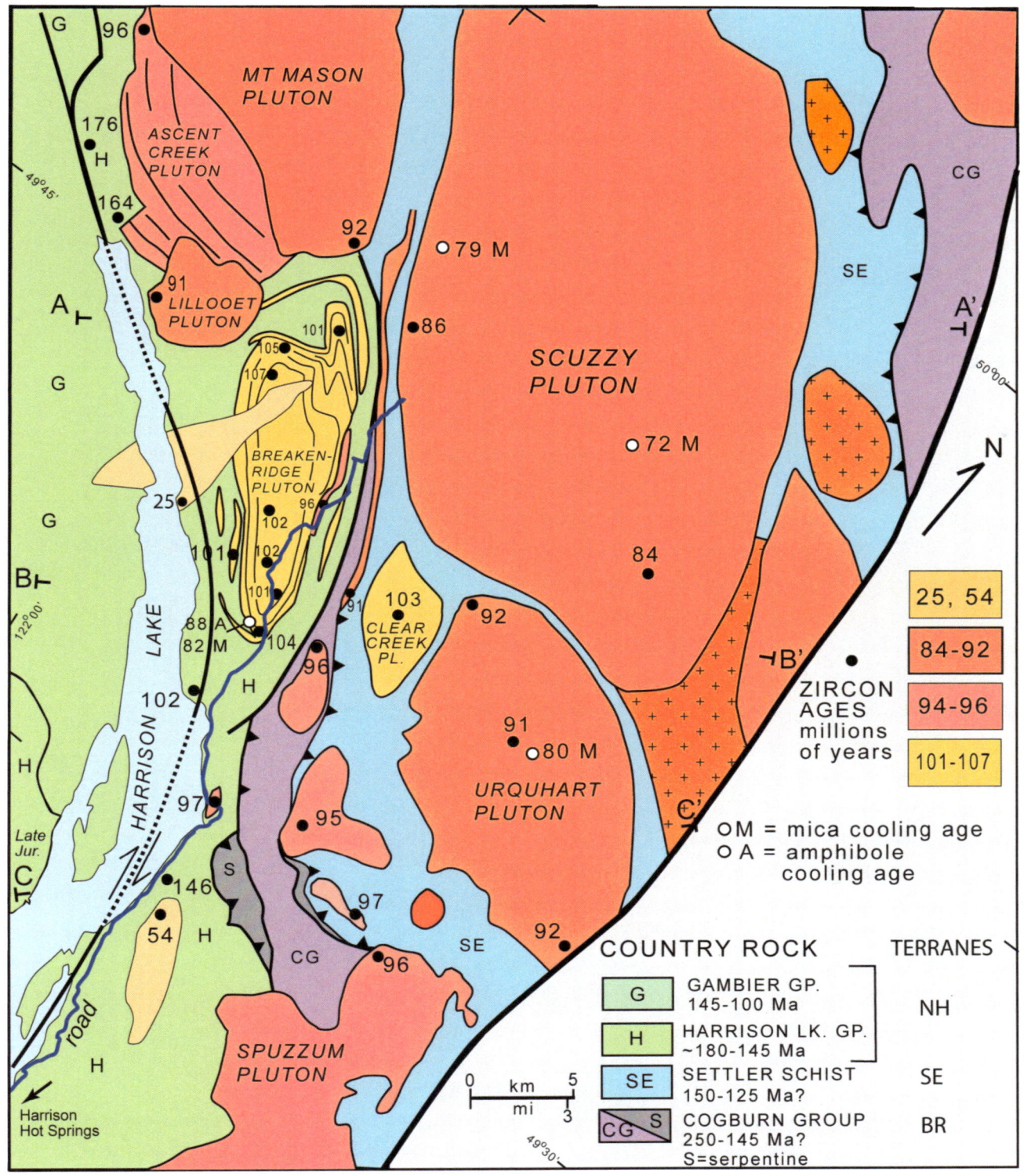

Fig. 11-4. Map of the bedrock geology of the Harrison Lake area. Pluton ages are for the most part based on U/Pb measurements in zircon (localities at the black dots), considered to be the igneous crystallization age. Mica and amphibole ages record the time when biotite mica cooled to about 300 degrees C, and hornblende cooled to ~550 C. Coupled with mineral evidence of burial depth, these cooling ages give us a valuable indication of when deeply buried rocks were uplifted.

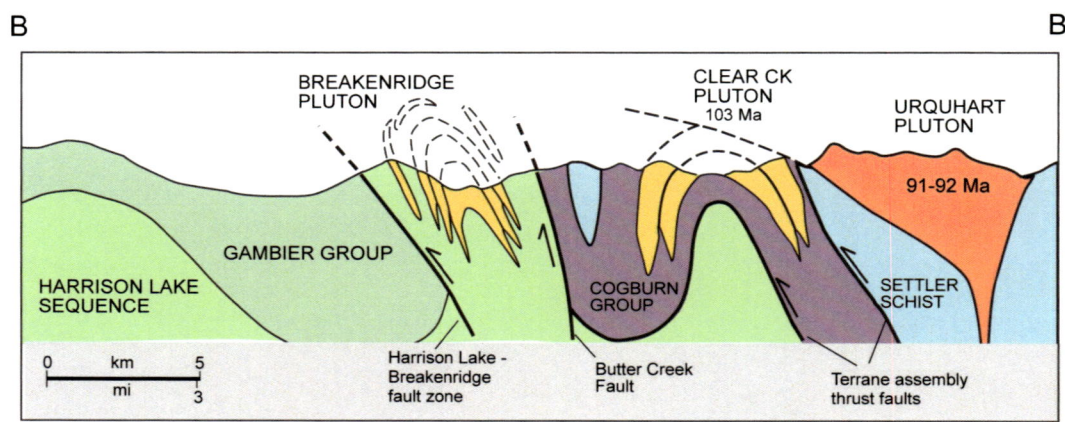

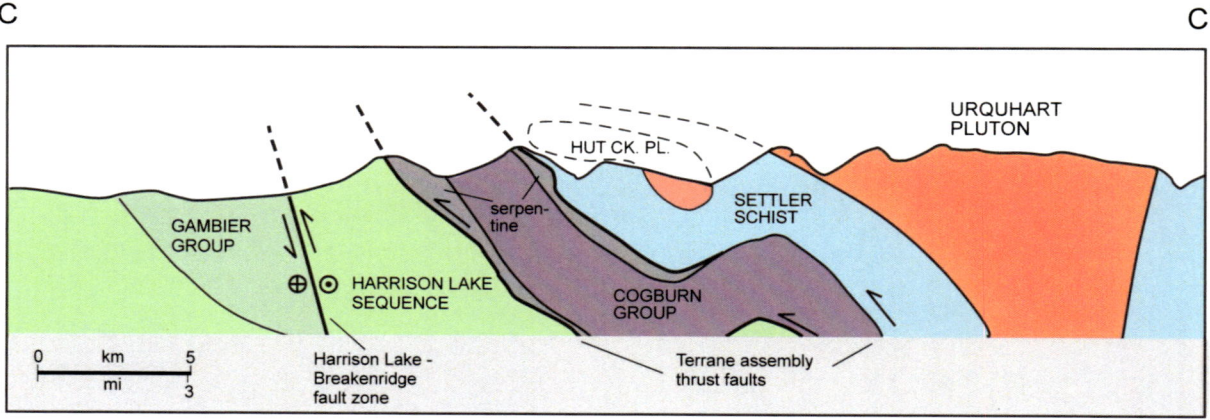

Fig. 11-5. Cross sections through the Harrison Lake area geology. Section locations on Fig. 11-4. Notes: 1) On A-A', the seismic reflection pattern showing layers in the country rock under at the Scuzzy pluton at its east side indicates a shallow west-dipping pluton floor. 2) Metamorphism and east-west contraction (folding) occurred after intrusion of the 101-107 Ma Breakenridge pluton, which is folded, and before the 91-92 Ma Urquhart pluton, 91 Ma Lillooet pluton, and 94 Ma Hut Creek pluton, all of which cross-cut the folds . 3) Dextral strike-slip faults (west side north), as well as reverse faulting, are indicated on the Harrison Lake - Breakenridge fault zone. 4) Terrane contacts between Settler - Cogburn, and Cogburn - Harrison are shown as thrust faults, but the evidence is scant—these contacts could be not much displaced from the original structure.

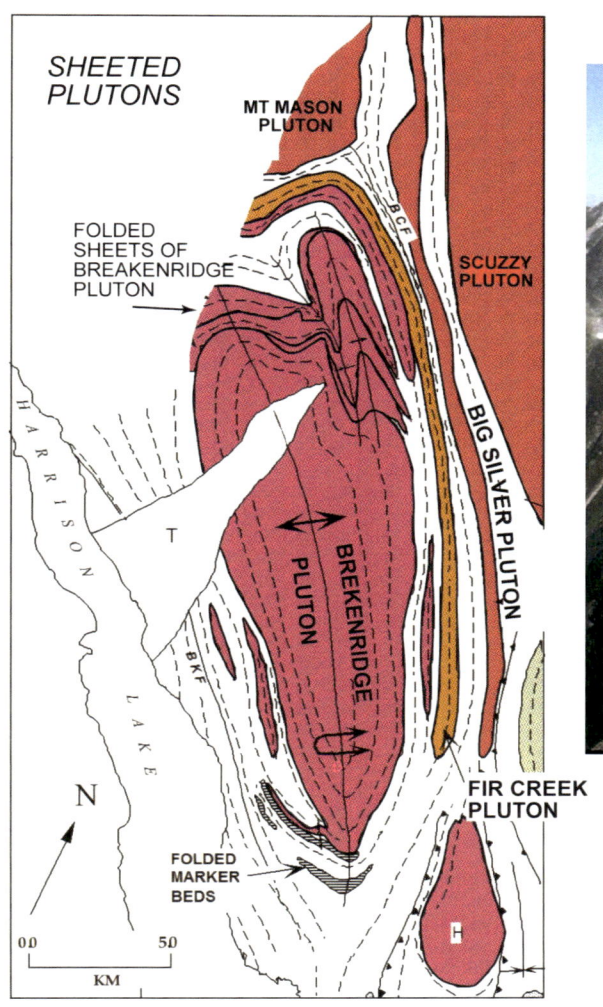

Fig. 11-6. Map of the Breakenridge pluton and other sheeted plutons in the Harrison Lake area.

Fig. 11-7. View north (from helicopter) showing topographic steps that define large-scale pluton sheets curving over along the northwest flank of the Breakenridge pluton. Peaks on the left skyline are in the Mt. Mason pluton.

Fig. 11-8. Breakenridge orthogneiss. The strong foliation is metamorphic, overprinted on the original igneous texture.

Fig. 11-9 Outcrop scale of light -colored pluton sheets intruded into darker gneissic country rock. Bernie Dougan high on Mt. Breakenridge.

Fig. 11-10. Pluton injection complex along the west flank of the Scuzzy pluton, in Shovel Creek, near the road up Big Silver Creek. Sills of quartz diorite are interlayered with country rock schist, suggesting a sheeted origin for the pluton as a whole. Ned for scale.

Fig. 11-11. View west of the Lillooet pluton from the Mt. Breakenridge area. This 91 Ma pluton, defined and named by Tom Lapen, is undeformed and is intruded across structural trends of the 101-107 Ma Breakenridge pluton and the 96 Ma Ascent Creek pluton. A floored structure to this pluton is suggested by inward-dipping igneous foliation low on the mountain, and also by an igneous cumulate zone at the base of the mountain.

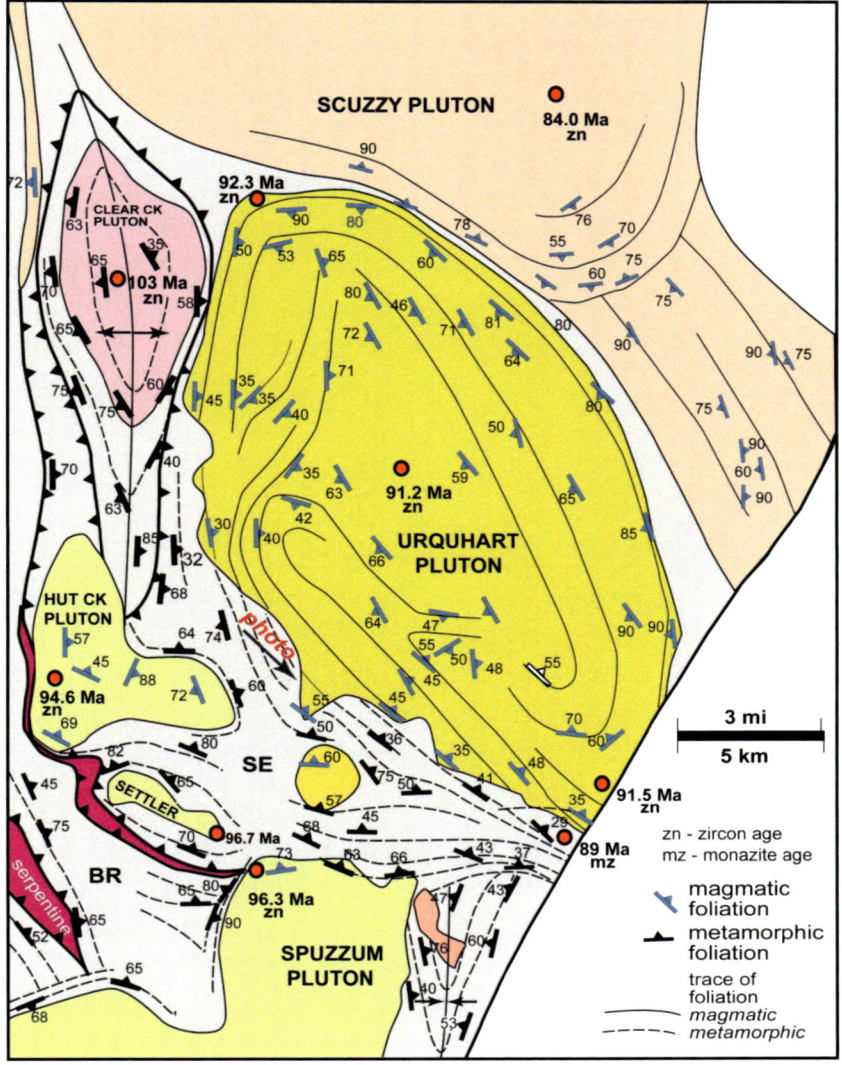

Fig. 11-12. Rock ages and fabrics in the vicinity of the Urquhart pluton. "Photo" in this figure marks the view shown in figure Fig. 11-13 below.

Plutons of the Harrison Lake area

How do the Harrison Lake plutons (Figs, 11-4, 11-5) fit into these emplacement models? The entire Breakenridge pluton is a sheeted complex. Sheets of quartz diorite were intruded between horizontal layers of country rock. Subsequently, the layered pile was folded into a doubly plunging anticline (Figs. 11-4, 11-5, 11-6, 11-7, 11-9). Individual layers can be recognized on an outcrop scale to regional scale. (11-20). The Breakenridge pluton formed as a stack of horizontal intrusive sheets.

The Breakenridge plutonic complex has been metamorphosed (Fig. 11-8). It has a strong planar fabric (foliation), and considering its origin as an igneous rock, the term orthogneiss is used. The metamorphic event predates the folding, as the gneiss is folded. The metamorphic foliation is parallel to the igneous layering, which was originally horizontal, and contains a strong lineation of mineral streaks that point to shear movement headed northwest. The ages of the metamorphic event, the northwest regional shearing, and the folding, are bracketed between the crystallization age of the Breakenridge plutonic rock ranging from 107-101 Ma, and the 91-92 Ma age of the Mt. Mason and Lillooet plutons that cross-cut the folds (Figs. 11-5, 11-11).

Fig. 11-13. View southeast of the floor of the Urquhart pluton resting on Settler Schist. See above figure for photo location.

Also showing evidence of an origin by sheeting is the Scuzzy pluton. At both the east and west margins of this giant pluton are wide zones (hundreds of feet or more) of pluton sheets interlayered with country rock (Fig. 11-10). At least in these marginal zones, the magma came in pulses. It would seem such a process is inconsistent with the mechanism of diaper intrusion, in which the pluton emplacement occurred by movement of a magma blob enmass through the country rock.

A partial depth profile of the Scuzzy pluton is evident from a seismic reflection study that defines a shallow inward-dipping floor overlying layered country rock on the east side of the pluton (Fig. 11-5 section A-A'). Was the whole pluton formed by a succession of horizontal sheets? Probably, but this is not proven because the internal geology of the pluton is relatively homogeneous and does not show distinct layering as in the Breakenridge pluton.

Fig. 11-14. *above:* **Sheeted margin along the west flank of the Urquhart pluton. Brownish country rock separates layers of pluton. View south.**

Next we look at the Urquhart pluton and surrounding Clear Ck, Hut Ck, Settler, and Spuzzum plutons, and country rock (Fig. 11-12). Here is a succession of plutons intruded over a time frame of 103-84 Ma. Are the older plutons and country rock shoved aside by later intrusions, indicating forceful space-making?

The Clear Creek pluton is the oldest at 103 Ma, and is a relative of the similar-aged metamorphosed Breakenridge pluton to the west. This pluton also has been metamorphosed, and the metamorphic layering is folded (Fig. 11-5 section B-B'). Emplacement of the Clear Creek pluton and its relatives in the Breakenridge pluton pre-date the metamorphism and folding.

Spuzzum pluton and nearby Hut Creek and Settler Creek plutons are 94-96 Ma in age. Urquhart pluton is 91-92 Ma. Scuzzy pluton is the youngest, 84-86 Ma. Looking at contacts of these plutons with each other and with the country rock gives us an indication of how the magma was emplaced.

Fig. 11-15. **Magmatic foliation within the western border area of the Urquhart pluton (outcrop near Bernie in Fig. 11-14); microscope view in Fig. 11-16. This fabric, here dipping east underneath the pluton body, forms as a mush of crystals and melt that is compressed and/or flows during pluton emplacement.**

The Spuzzum pluton has thin aureole of strain in the country rock—varying from less than measurable to ~ 300 meters (Fig. 11-12). Some flow of country rock occurred locally around the intruding mass.

The Urquhart pluton shows a gently dipping floor on the south side (Fig. 11-13) and an inward dipping sheeted margin on the west side (Fig. 11-14). Magmatic foliation within the Urquhart pluton (Figs. 11-12, 11-15, 11-16) marks flowage patterns of crystallizing magma, defining a roughly tub-shaped body. The country rock foliation in the Settler Schist defines a relatively broad zone that follows

Figs. 11-16 to 11-19. Typical mineralogy and textures of plutons in the Harrison Lake area. Many plutons show a discernable alignment of minerals defining a foliation created by magmatic flow. The dominant minerals throughout are quartz, plagioclase feldspar, hornblende, and biotite. Plutonic rocks are mostly clear of country rock inclusions, but do have autoliths, dark hornblende-rich rocks derived from basalt intermingling at depth in the granitic magma chamber.

Fig. 11-16. *left:* Magmatic foliation viewed with the microscope. The dominant grains, grey to white, are laths of plagioclase feldspar. The colored grains are biotite and hornblende. Alignment of the grains happens as crystallization occurs in flowing magma. The color of the grains is partly a function of the polarized light. Coming upward from the base of the microscope, light is polarized east-west; light having passed through the grains goes through a north-south polarizer. Crystals interact to this light with their own polarizing effects to give colors.

Fig. 11-17. *right:* Close-up of the Urquhart pluton. Dark minerals are mostly hornblende. Light minerals quartz and feldspar. The very dark inclusion, termed an autolith, is likely a quenched blob of basalt magma injected into the much cooler granitic magma.

Fig. 11-18. *below*: Cut slab of Scuzzy pluton stained with chemicals that color feldspar (pink) but not quartz, allowing for an accurate measure of the proportions of these two minerals.

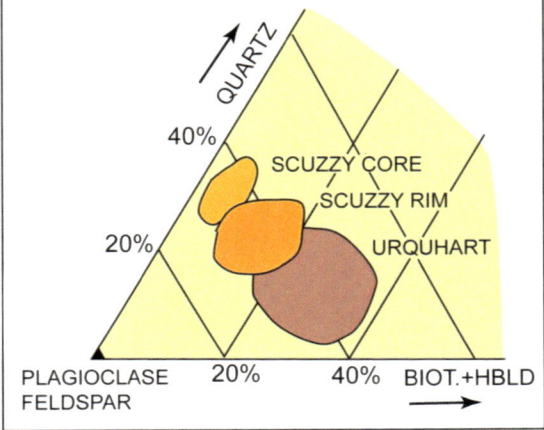

Fig. 11-19. *above.* Proportions of minerals in plutons measured by "point-counting" with the microscope on stained slabs.

97

the pattern of magmatic foliation at the south end of the pluton (Fig. 11-12), but is narrow along the western flank.

Is the country rock foliation around the Urquhart pluton an indication of plastic flow around a diapir; or is it a consequence of compaction under the weight of an overlying pluton that has grown by piling up of horizontal sheets? Evidence of the latter mechanism is 1) the absence of discernable displacement of the margin of the older Clear Creek pluton lying just to the west, and 2) a sheeted margin on the west flank of the pluton (Fig. 11-14).

The Scuzzy pluton shows no displacement of the older, adjacent, Urquhart pluton. Remarkably, there is no apparent strain margin for the Scuzzy pluton. We can recollect from Fig. 11-10 that the Scuzzy shows evidence of sheeting along its margin; this, and the lack of country rock strain along the margin, implies a passive emplacement and supports the concept of pluton growth by horizontal sheeting. We will revisit the question of pluton emplacement with insights from metamorphism of the country rock.

In summary, pluton emplacement by *Stoping* is ruled out by a near absence of country rock inclusions. *Diapirism* could not have been a major mechanism, considering the minimal disruption of country rock at pluton margins. Likewise, absence of significant lateral displacement of country rock precludes *vertical sheeting*. *Horizontal sheeting* is supported by: 1) observed layering, 2) low-dip pluton floors, and 3) minimal lateral displacement of country rock. Sheet stacking is inferred to have occurred by successive pulses of magma rising from depth, spreading laterally, and growing a thick pile. Sheets could have been added at either the bottom or top of the growing stack. Metamorphic evidence that the bottom was pushed down suggests that the additions were at the top.

Fig. 11-21 Highly smeared-out conglomerate of the Nooksack-Harrison terrane. Pebbles and cobbles are arc derived.

Fig. 11-20 View southwest of Mt. Breakenridge. Linear bands of snow follow the sheeted structure in the Breakenridge pluton.

Fig. 11-22. Ribbon chert of the Cogburn Group (= Bridge River terrane). White quartz-rich layers alternate with fine seams of mica schist. The rock originated as a deep sea ooze composed mostly of siliceous plankton—radiolaria. Metamorphism has deformed the layers, but the original sedimentary structure is preserved.

Country Rock

In our quest to understand the dynamics of mountain building, features of the country rock, especially metamorphic, are critical.

Three country rock units are present in the Harrison Lake area: Nooksack-Harrison, Cogburn, and Settler (Figs. 4-2, 11-4).

The Nooksack-Harrison terrane includes both the Harrison Lake and Gambier groups of Fig. 11-4. These rocks all have a strong flavor of arc origin, both as volcanic rock and sedimentary derivatives (Fig. 11-21). As described in Chapter 4, the Nooksack-Harrison terrane formed in a back-arc basin that extended to Alaska, and may have moved south some hundreds of miles relative to neighboring terranes on the east.

The Cogburn Group we equate to the regional Bridge River terrane, a long-lived fragment of ocean crust (Figs. 4-3, 4-4, 4-5). The Cogburn in our area of interest shows all levels of an oceanic plate, albeit metamorphosed: a serpentinite

Fig. 11-23. *left:* Settler Schist, a strongly metamorphosed shale. This unit is a likely derivative of the forearc muds and sands of the Methow Basin (Fig. 4-5). Coarse white lenses are sillimanite grains crystallized in replacement of contact metamorphic andalusite (Fig. 3-9). This outcrop is in the contact aureole of the Urquhart pluton.

component represents the top of the mantle; amphibolite, derived from basaltic crust of ocean-ridge origin; and deformed but still recognizable ribbon chert (Fig. 11-22), originated as deep ocean radiolarian ooze.

The Settler Schist (Fig. 11-23) is, across the region, a close companion of the Bridge River terrane. It is derived from mud and sand. As described in Chapter 4, studies suggest correlation of the Settler Schist with forearc sediments of the Methow basin, early Cretaceous in age (100-160 Ma), and underlain by the Bridge River ocean crust. In the Harrison Lake area, the Settler Schist overlies the Cogburn/Bridge River unit. The contact has been mapped as a thrust. Certainly there is deformation along the contact but perhaps not so much fault displacement if we correlate these metamorphic rocks to the Methow sequence.

Metamorphism

Plutonism and metamorphism have greatly altered the country rock units. Plutonism we have looked at above, now we'll investigate the metamorphism. Here's a Google definition of "metamorphosis" I like that includes what happens to a rock: "*a change of the form or nature of a thing or person into a completely different one, by natural or supernatural means*".

For rocks, the agents of metamorphism are temperature, confining pressure, and shear pressure. Chapter 3 outlines the theory and the important minerals for evaluating metamorphic pressure - temperature conditions in rocks of interest. Notably, the pressure - temperature gradient in the San Juan Islands - northwest Cascades thrust system is mainly that of temperature

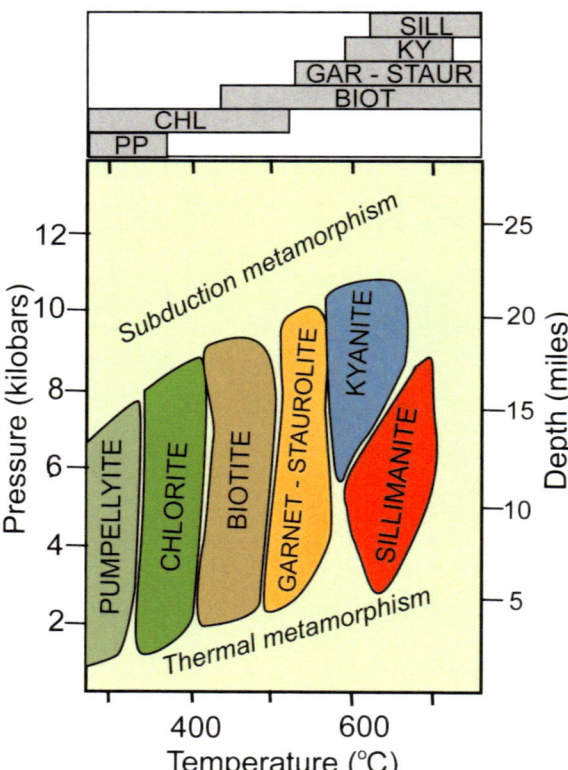

Fig. 11-24. "Barrovian" metamorphic zones. The upper panel shows the temperature range of each index mineral.

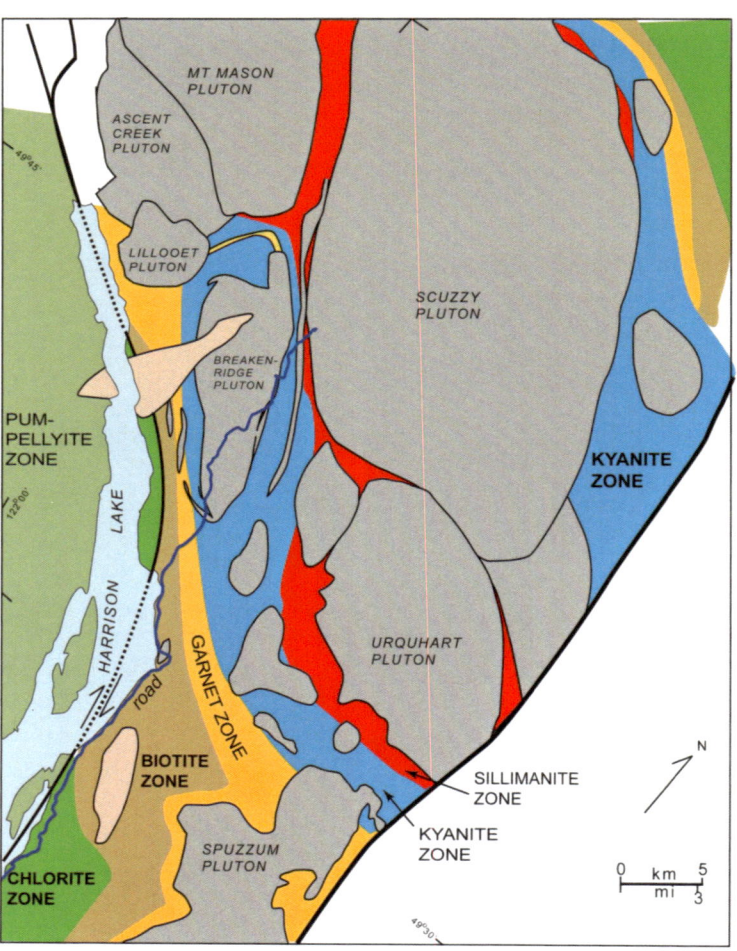

Fig. 11-25. Metamorphic zones in the Harrison Lake area based on index minerals. The pattern is that of Barrovian metamorphism.

variation, whereas that of the Coast Plutonic Complex is mainly a pressure gradient (Fig. 3-15). We'll come back to this.

Country rocks of the Coast Plutonic Complex in the Harrison Lake area exhibit a classic "Barrovian" metamorphic sequence (Figs. 11-24, 11-25) caused by crustal thickening in the roots of a mountain range. George Barrow developed this concept more than 100 years ago in the Scottish Highlands. At

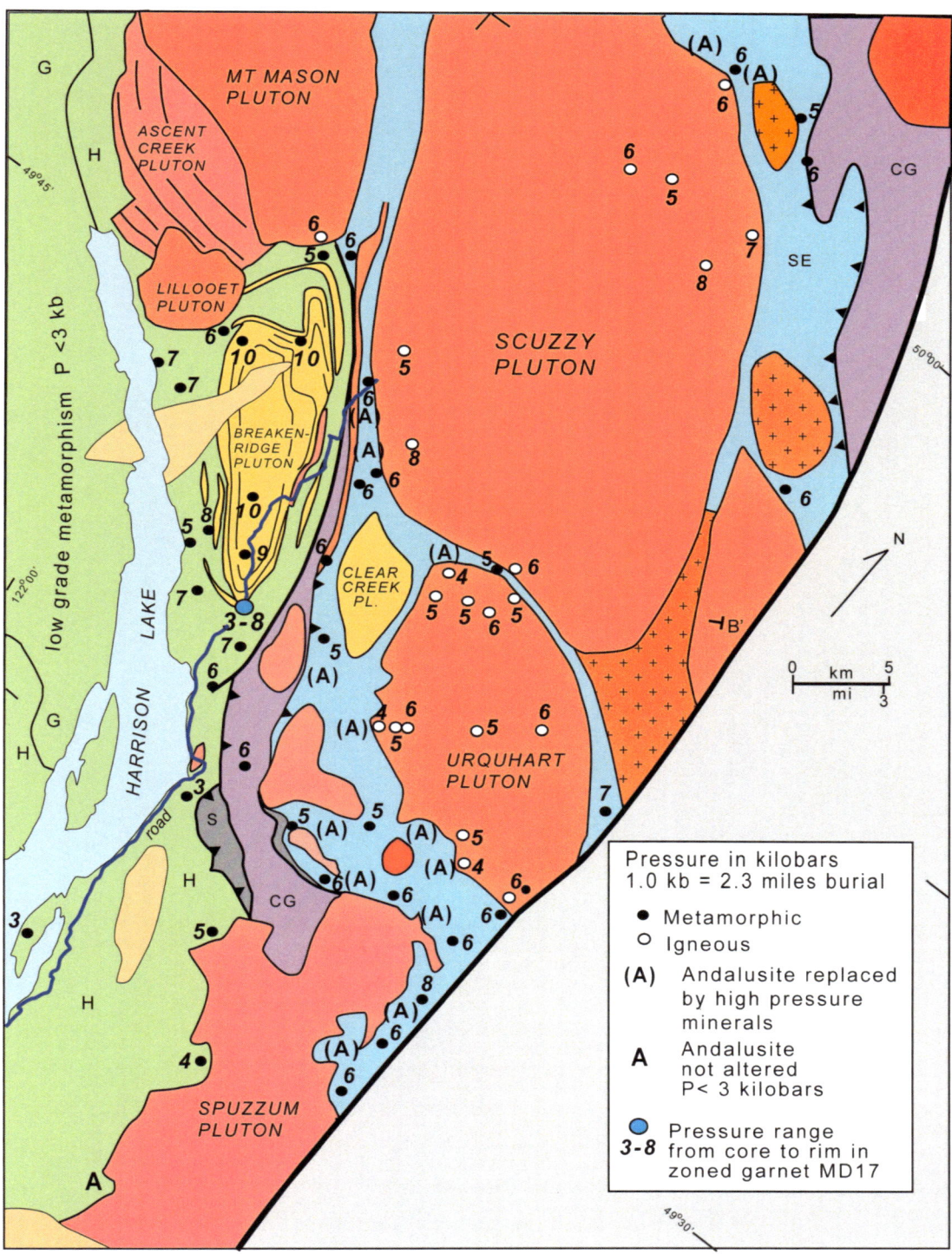

Fig. 11-26. Map of metamorphic pressures in country rock and igneous pressures in plutons of the Harrison Lake area.

101

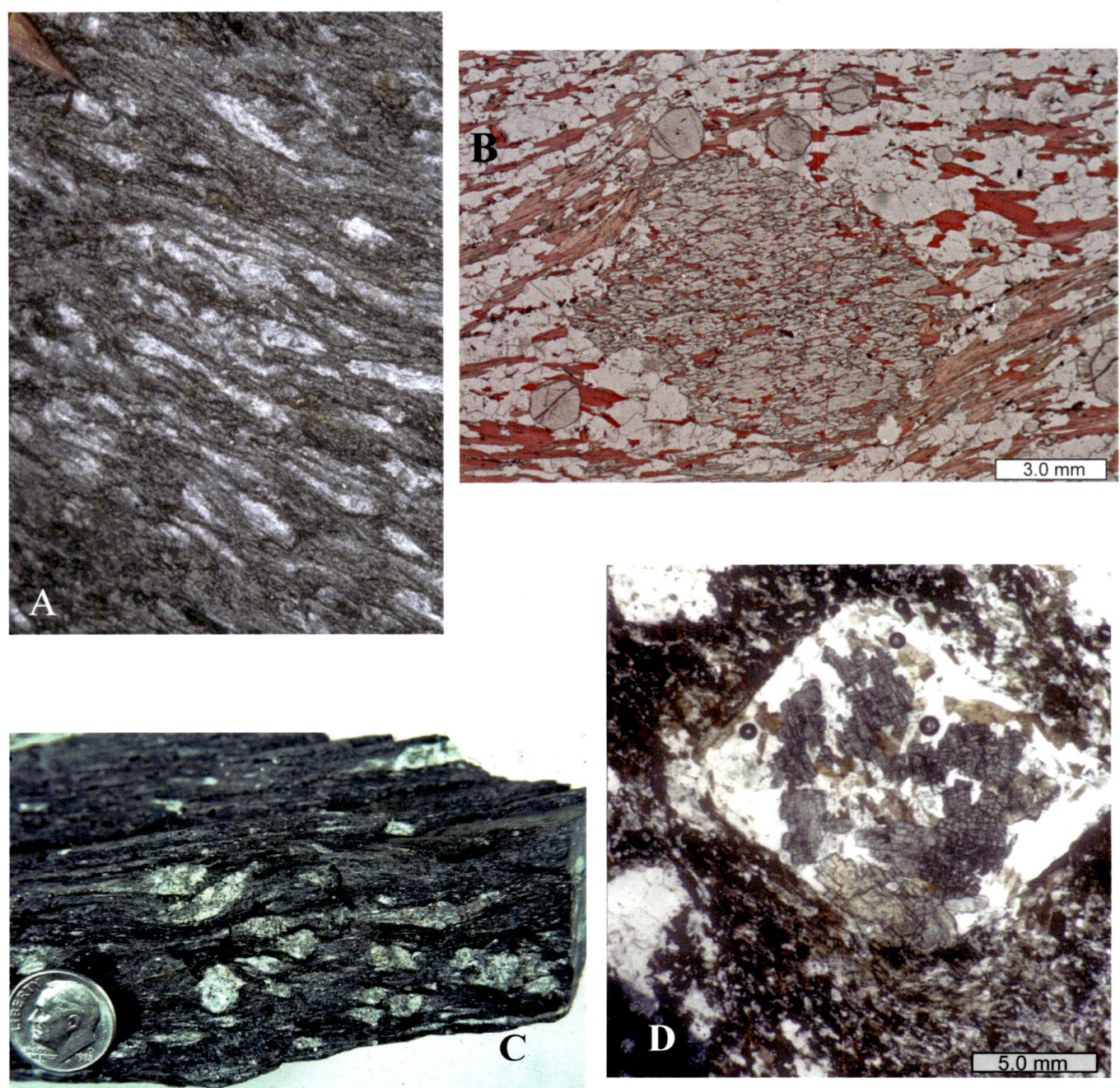

Fig. 11-27. Andalusite laths replaced by kyanite and sillimanite, Settler Schist, Harrison Lake area. **A.** Outcrop view. White laths are bundles of sillimanite prisms aligned parallel to the elongate grains of original andalusite that they replace. **B.** Microscope view of a cross section of pseudomorphed andalusite. Countless fine prisms of sillimanite, aligned with the original andalusite, show flattened diamond shape in cross section. Biotite and garnet are also present. **C.** Cut hand sample of schist with outlines of deformed andalusite grains, now replaced by kyanite and staurolite. **D.** Microscope view of a cross section of original andalusite that is replaced by kyanite, staurolite, quartz, and feldspar.

that time he recognized a sequence of index minerals as marking, in a qualitative way, an increase in pressure and temperature that indicate the degree of metamorphism. Decades later, quantitative values of temperature and pressure became known through laboratory experiments of mineral stability. The Barrovian index minerals are notably sensitive to temperature (Fig. 11-24), but not so much to pressure variation. For our purposes, the Barrovian zones certainly mark the focus of orogeny in terms of temperature—around the plutons (Fig. 11-25).

To quantify pressure we need other means, as explained in Chapter 3. For country rock, useful barometers and thermometers are found in compositions of garnet, biotite, feldspar (plagioclase), and the presence of aluminum silicate polymorphs—andalusite, kyanite, and sillimanite. For plutons, we have a barometer in the aluminum content of hornblende (also discussed in Chapter 3). These calculated pressures are plotted on Fig. 11-26.

Also shown on Fig. 11-26 are known occurrences of relict andalusite replaced by sillimanite or kyanite. This pseudomorphed andalusite is a key player in our analysis of regional tectonism. It marks an early, hot (~600°C) metamorphic event at low pressure, depth less than 4 to 5 miles (Fig. 3-15). Additional evidence of early low-pressure metamorphism comes from the low calcium content of the core of a zoned garnet on the flank of the Breakenridge pluton (Figs. 3-13, 11-26). Most everywhere in country rock marginal to plutons, we see the relict andalusite overprinted by high-pressure minerals. This tells us that country rock marginal to plutons was initially relatively close to the earth's surface, less than 3 to 5 miles. Subsequently, the rocks were deeply buried.

Plutonic and Metamorphic Interplay
What caused the metamorphism of country rock? Certainly from spatial association of high grade metamorphism with the Scuzzy, Urquhart, and Breakenridge plutons (Fig. 11-25), we could speculate a causal relationship between the plutonic and metamorphic events. One thought could be that metamorphism melted local rock to make the plutons. But, for many reasons, the pluton magma is regarded as originating near the base of the earth's crust, relating to influx of magma from a subduction zone (Fig. 1-6).

Deep burial of sedimentary rocks marked by Barrovian metamorphism is commonly found in the roots of mountain belts around the globe. In most places, the cause is reasonably attributed to thickening of the earth's crust by contraction involving folding and/or faulting—typically the emplacement of thrust sheets (Alps, Fig. 2-2). Could this structural explanation apply to the origin of high-grade metamorphic rocks of the Harrison Lake area? We do have a regional structural marker that indicates contraction: the planar contact between the Cogburn and Settler terranes. There is a significant fold of this surface that thickens the crust, but it predates the Scuzzy and Urquhart plutons, and postdates the Breakenridge pluton (Fig. 11-5). No other regional folds are evident that would have thickened the crust during the Barrovian metamorphism.

What about the possibility of a giant thrust sheet that came across the Scuzzy and Urquhart plutons? Several problems come to fore: 1) There is no evidence of a regional root zone outside the pluton area from which the thrust sheet could have risen. 2) The thrust would need to come across the plutons *during* their emplacement, and since the two plutons were built at different times (86-84, 92-91 Ma), we would need two thrusting events. Further, if there were two thrusts, the first would somehow have been absent from the country rock of the Scuzzy pluton, which was still at a shallow level when initial phases of the Scuzzy were intruded (andalusite aureole). 3) Thrusts leave outcrop-scale markers as strain features, such as shear zones and foliation. We don't find these in the Scuzzy and Urquhart plutons, only igneous fabrics. 4) As far as we know, the Coast Plutonic Complex from Harrison Lake west to Wrangellia bears no evidence of high-pressure metamorphism (Fig. 11-26) or great thrusts.

We could then turn to the idea that somehow the Scuzzy and Urquhart plutons caused the

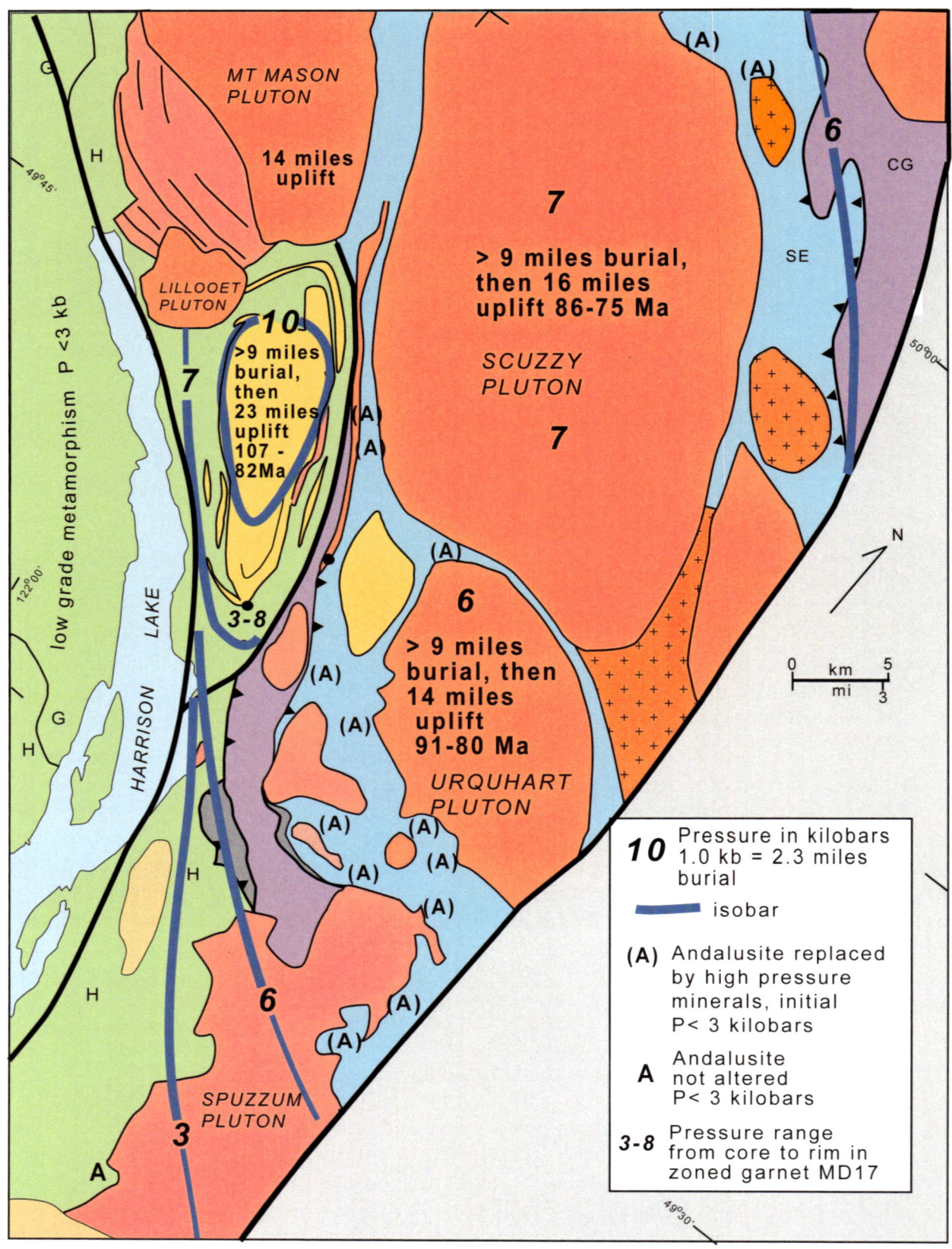

Fig. 11-28. Summary of metamorphic pressures and correlative depths of burial and uplift for the Harrison Lake area. Age ranges (e.g. 107-82 Ma) bracket the depth history sequence of shallow-buried-shallow, as best constrained from available ages.

metamorphism in their vicinity. This makes sense in that plutons bring a vast amount of heat into the upper crust. "Contact metamorphism" is well known. A complication arises, however, in that the country rock is not only heated in proximity to plutons, but it also becomes deeply buried. Away from these big plutons, metamorphic minerals formed at relatively low pressure, indicating burial of less than five miles. In the vicinity of the plutons, country rock records high pressures (Fig. 11-26); burial was in the range of 15 to 23 miles.

What went on with these plutons? The plutons are layered, and they have floors. The plutons appear to be tub-shaped from what we can see from their inward dipping margins and patterns of igneous foliation within the bodies. The plutons started out as shallow bodies, indicated by the andalusite (Figs. 11-26, 11-27). Further intrusions into the same pluton body crystallized at deeper levels, as documented from the aluminum content of hornblende crystals. The plutons thus thickened greatly as they formed. Following this hypothesis, the pluton floor and the underlying country rock were apparently pushed down. When magmatism ceased, isostatic rebound elevated plutons and country rock, and much erosion took place at the surface, exposing the deep levels we see now.

For the Breakenridge pluton, an origin of the pluton by stacking of horizontal sheets, spread over a time frame of 107-101 Ma, is well documented. The pluton layers are defined by interleaving with the country rock (Figs. 11-6, 11-7, 11-9). Both pluton and country rock are metamorphosed, the pluton being a well-foliated gneiss (Fig. 11-8). The gneissic fabric is folded. All this happened after the youngest age of the Breakenridge pluton, 101 Ma, and before the 91-92 Ma ages of the Lillooet and Mt. Mason plutons, which cross-cut the metamorphic fabric and the folds.

Metamorphism of country rock in the contact aureole of the Breakenridge pluton is documented in zoned garnet to have ranged in depth of burial from less than 5 miles for the garnet core to 20 miles for the rim (Figs. 3-13, 3-14, 11-26, 11-28). Pressure in the aureole apparently increased during plutonism. Additionally we see, from the distribution of metamorphic pressures in the region, that the loading was centered on the Breakenridge pluton (Figs. 11-26, 11-28). Magma loading seems probable.

Mechanics of Arc Magmatism
Let's look at the Coast Plutonic Complex in the Harrison Lake area in the broad context of how arcs in general form.

The question of how magmatic arcs are built has given a cadre of geoscientists much to chew on. I once attended a four-day conference in Valdez, Alaska , where ~60 scientists from all parts of the world discussed this issue. Here is a summary of themes of the conference, including my own presentation of the "magma loading" concept as deduced from the Harrison Lake area.

As presented in Chapter 1 (Fig. 1-6), we view the process of arc formation to start at an ocean ridge where mantle peridotite is hydrated to serpentine and chlorite. The ocean crust, including serpentine and chlorite rock, travels to the subduction zone. There, the chlorite and serpentine, carried down to great pressure, react, releasing water and reverting back to peridotite. The water rises from the subduction zone and initiates partial melting of mantle material in the overriding plate (Fig. 11-29); a magma of basalt composition is produced. The basalt magma rises into the lower crust. Here it begins to crystallize, first forming pyroxene and olivine crystals. These crystals are heavier than the remaining melt and they settle, forming a peridotite cumulate on the floor of the magma chamber. At the same time, surrounding country rock of the lower crust is heated by the incoming magma and partially melts, especially the feldspar and quartz components. By these means of settling-out of some minerals and melting of others, the original basalt melt becomes a granitic magma. This magma rises, creating a pluton and volcanic eruptions. The pyroxene and olivine cumulate is left behind.

The crust thickens and the Moho sags; compensation for the thickened crust happens by downwarping in the plastic asthenosphere. Confining pressure rises at the base of the crust owing to the increased load above, and reaches the point where pyroxene in the cumulate converts to garnet, a more compact mineral composed of the same elements. More compact equates to higher density, so the cumulate rock, now composed of garnet and olivine, is heavy enough to break away and sink down from the Moho deep into the mantle.

The granitic magma, separated from the cumulate streams upward in episodic pulses, forming a pluton higher in the crust. In the vicinity of granitic plutons intruded at shallow levels in the crust, country rock is affected by contact metamorphism. Andalusite crystallizes, marking a depth less than 5 miles. Magma pulses continue for a few million years, thickening the pluton by addition of sheets. The original andalusite is pushed down, eventually reaching the higher pressure stability field of kyanite and/or sillimanite.

With the cessation of magmatism, the thick granitic blob that is the pluton has greatly thickened the crust. The density of this added material is relatively low. We've discussed in Chapter 1 the concept of the "floating equilibrium" between earth's lithosphere and under-lying plastic asthenosphere. The relatively light-weight thickened crust is like a submerged cork in water—it floats up. Given time, erosion removes the mountain mass that rises, and the deep levels of the pluton and country rock margin are exposed for geologic analysis and speculation.

Fig. 11-29. Cartoon diagrams of hypothesized stages in the formation and uplift of a pluton in the Harrison Lake area. A = andalusite, K= Kyanite.

Places to see Harrison Lake Rocks

An excellent field guide to the Harrison Lake geology described in this chapter is available online. In the references section, see a listing for Gibson and Monger, 2014.

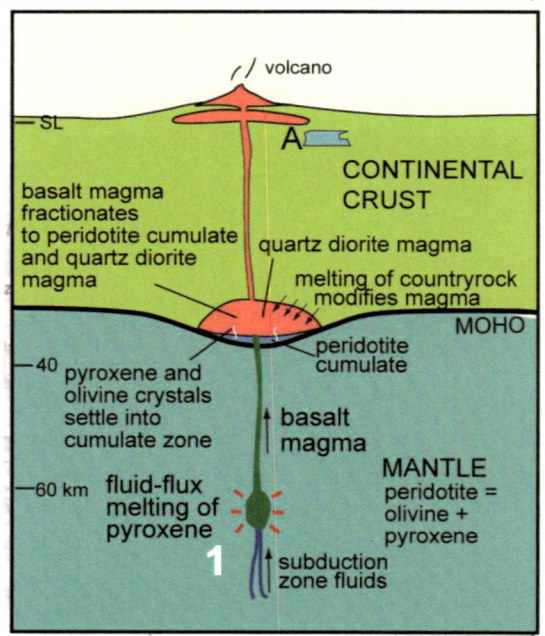

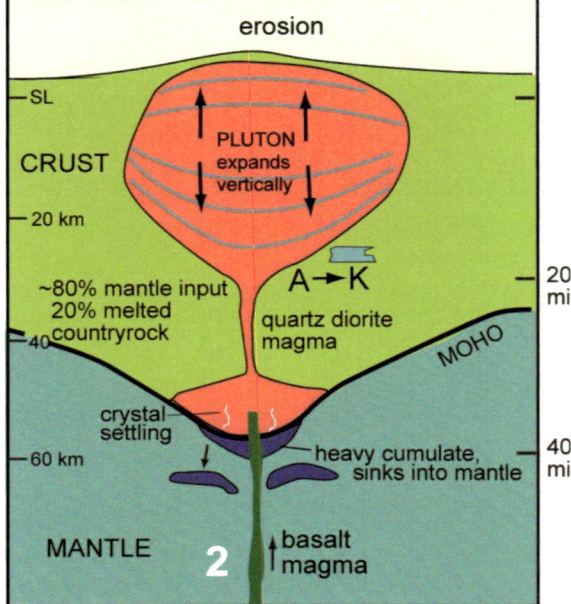

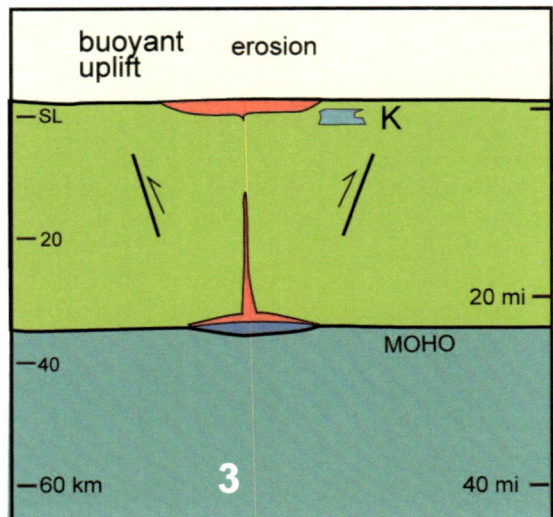

CHAPTER 12

CASCADE CRYSTALLINE CORE

Fig. 12-1 Mountains and glaciers of the North Cascades. View east from Eldorado Peak.

Introduction

The Cascade Crystalline Core (Cascade Core) is a land of granitic mountains, glaciers, and wilderness . Much of this landscape lies in the North Cascades National Park (Figs. 12-1, 12-2). The country, known as the "American Alps", attracts adventurers from afar. I was one back in the mid-1960s, when as a graduate student at U.C. Berkeley, I headed up to the Cascades for a break, and some mountaineering.

This trip led to my dream job at Western Washington University and a lifetime of exploration and research in this extraordinary country.

The Cascades Core constitutes the south end of the 1000-mile-long Coast Plutonic Complex, which, as we have seen in the Harrison Lake area (Chapter 11), is a mix of plutons and host country rock terranes. With restoration of the Straight Creek - Fraser River fault (Fig 4-3), the Cascade Core is seen as an extension of the Harrison Lake geology.

Fig. 12-2 Ned on Magic Mountain with view of Eldorado Peak, 1965.

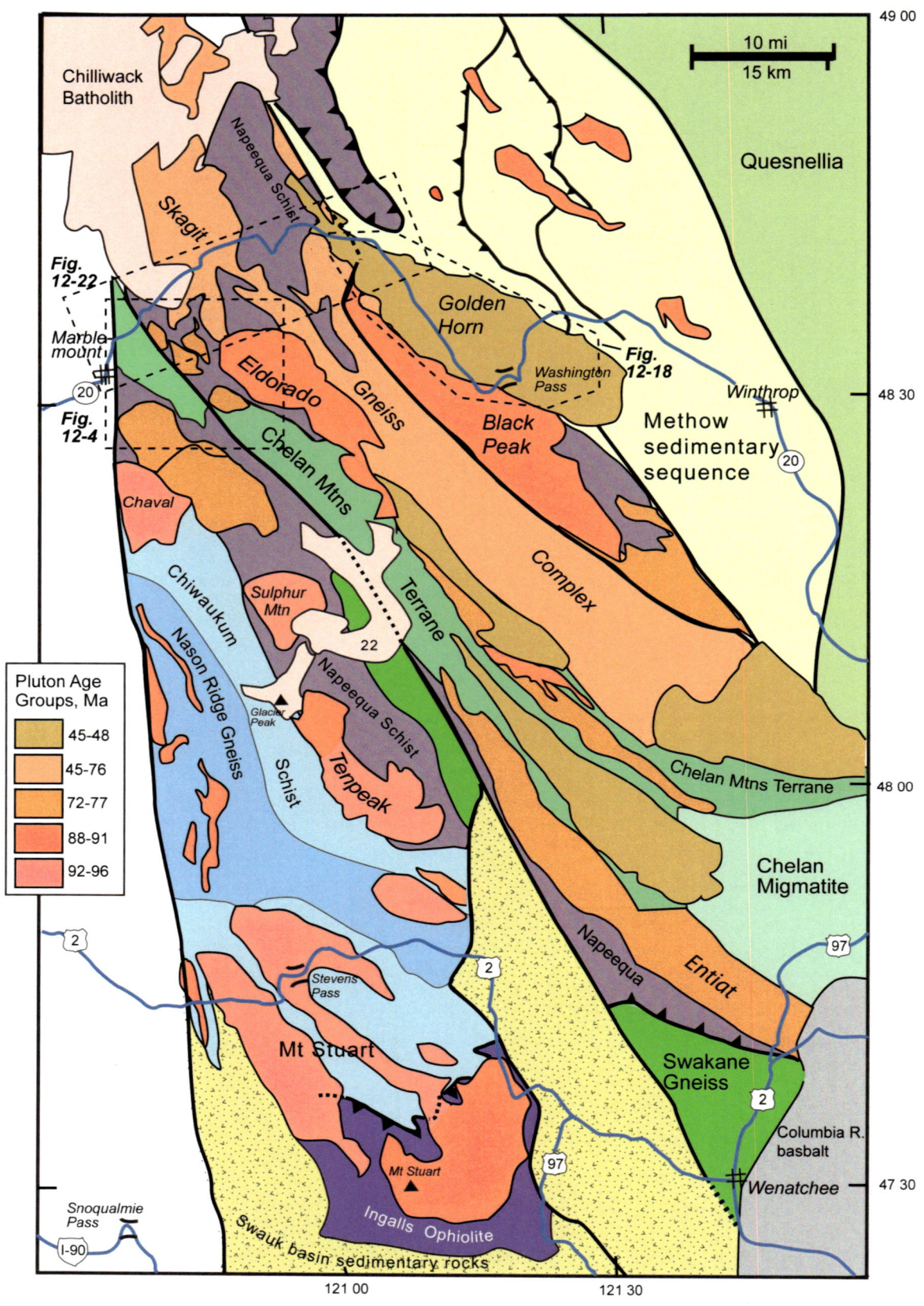

Fig. 12-3 Bedrock geologic map of the Cascade Crystalline Core. Dashed lines enclose areas of subsequent detailed maps.

The challenge of unravelling the mountain-building processes creating the Cascades Core has attracted much geologic attention. How do we explain two notable processes of this orogen: 1) the deep burial of once-surficial rocks, and 2) the emplacement of great amounts of granitic magma? These questions addressed for the Harrison Lake area were, in my reasoning, solved by a magma loading mechanism— the granitic plutons spread out in sheets piling up in great thickness (~20 miles) over the country rock. Does this mechanism apply also to the Core? Or, is there evidence of great crustal thickening by thrusting or other tectonic process?

Country Rock of the Cascade Core

We start with a consideration of the country rock. Mostly, these rocks are the same terranes as in the Harrison Lake area: The Settler Schist terrane at Harrison Lake = Chiwaukum Schist terrane in the Cascades, Cadwallader = Chelan Mountains terrane, Cogburn = Napeequa = Bridge River terrane (Fig. 4-3). As discussed in Chapter 4 (Fig. 4-5), a possible reconstruction of these country rock terranes puts the Bridge River at the base, the Cadwallader /Chelan Mountains terrane as an arc in the Bridge River ocean, and the Settler/Chiwaukum terrane a forearc sedimentary deposit eroded from the continental margin in early Cretaceous time (~120 Ma). *Note:* This designation of terranes departs from that of Tabor et al. 2003, who define the Chelan Mountains terrane as including the ocean floor (Napeequa) rocks with the arc rocks (Cascade River Schist and plutons). Here we separate the ocean floor and arc rocks, to be correlative with the B.C. scheme.

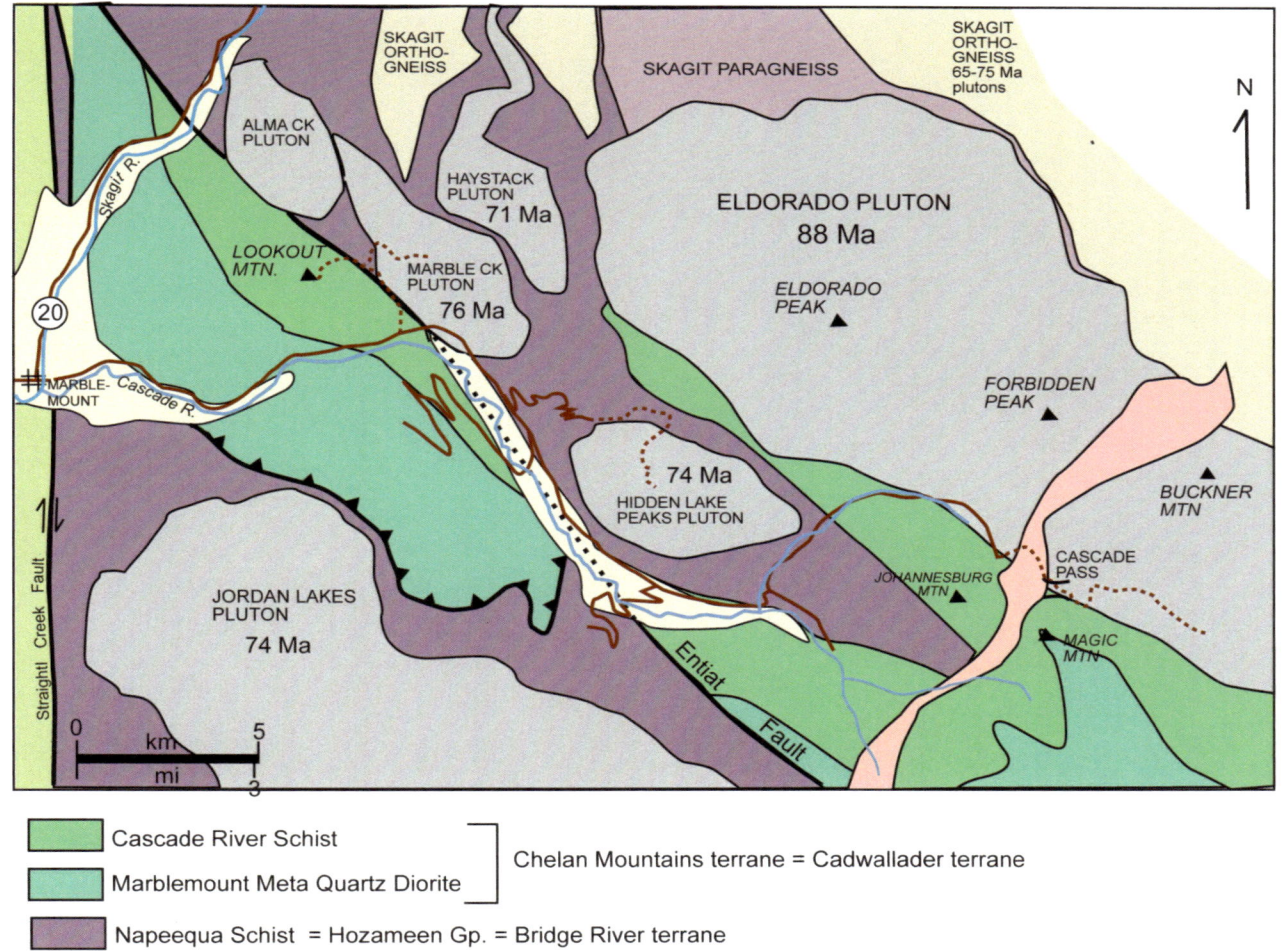

Fig. 12-4. Geology in the Cascade River area; map locality marked on Fig. 12-3.

In these country rock terranes, we are particularly looking for structural relationships, ages of rock origin and time of burial, and metamorphic temperature and pressure. All are features that could indicate tectonic history.

A very good place to get started on Crystalline Core geology is in the Cascade River area of the Chelan Mountains terrane (Figs. 12-4, 12-5). This is a well-studied region, both by Rowland Tabor of the USGS and by faculty and M.S. students at WWU. It is also an area of great attraction to mountaineers.

Fig. 12-5. Cascade River Schist. *Top left*: **View north of Lookout Mountain and outcrops of Cascade River Schist. On the horizon in background are glacier-clad Mt. Baker on the left, and Mt. Shuksan on the right.** *Top right:* **Close-up view of quartz diorite cobbles eroded from an arc plutonic rock, the "Marblemount Meta-Quartz Diorite" underlying the sediments.** *Bottom:* **Outcrop of interbedded sandstone and conglomerate near the summit of Lookout Mt.**

We'll start with a hike up Lookout Mountain (Fig. 12-5). Rock here is a relatively less metamorphosed part of the Cascade River Schist. Outcrop shows interlayered sandstone and conglomerate. And, we see evidence of plastic deformation related to metamorphism in the form of flattening and folding of the sedimentary rocks. The Cascade River schist is underlain by volcanic rock, which grades down into plutonic rock, named the Marblemount Meta Quartz Diorite. Both the volcanic rock and diorite give zircon ages of 220 Ma. This rock assemblage apparently represents a top-to-bottom section of an arc volcano tipped up on edge and greatly eroded.

Napeequa Schist commonly occurs near the Chelan Mountains terrane. The rock is metamorphosed ocean crust, consisting of amphibolite derived from ocean-floor basalt, metamorphosed mantle rock now mostly talc deposits, and banded quartzite originating as ribbon chert (Fig. 12-6). Contacts between the Napeequa and Chelan Mountains terrane mostly are interpreted as faults; but some may be depositional, inherited from the ocean floor setting.

Chiwaukum Schist (Figs. 12-7, 12-8) is mostly derived from shale and silt and thus has the right aluminum-rich chemistry for the growth of Barrovian metamorphic index minerals that show up in outcrop as large crystals (porphyroblasts) of andalusite, kyanite, sillimanite, and staurolite.

Fig. 12-6. Metamorphosed and deformed ribbon chert of the Napeequa Schist. The chert is a deep sea deposit, formed far from land. Burial, and consequent heat and pressure leading to metamorphism, weakens and folds the rock. Compare with the Cogburn terrane, Fig. 11-22. Photo from Haugerud and Tabor (2012).

Fig. 12-7. Chiwaukum Schist along the banks of Icicle Creek near Chatter Creek camping area. Original sedimentary bedding of siltstone and shale is preserved in this rock, but is strongly deformed. This is a relatively low-grade metamorphic rock, lacking coarse metamorphic grains as appear in this unit farther north, seen in the next figures (Fig. 12-8). Note the tight fold in the center of the photo—the fold axis is close to the outcrop surface.

Fig.12-8. *Top*: Outcrop of Chiwaukum Schist along the Wenatchee River. White layers are quartz-rich segregations formed during metamorphism. *Below left*: Detail of outcrop above. White kyanite laths are aligned. Compare with the Settler Schist, Fig. 11-23. *Below right*: A different outcrop: quartz-rich Chiwaukum Schist in which quartz is dispersed throughout the rock.

112

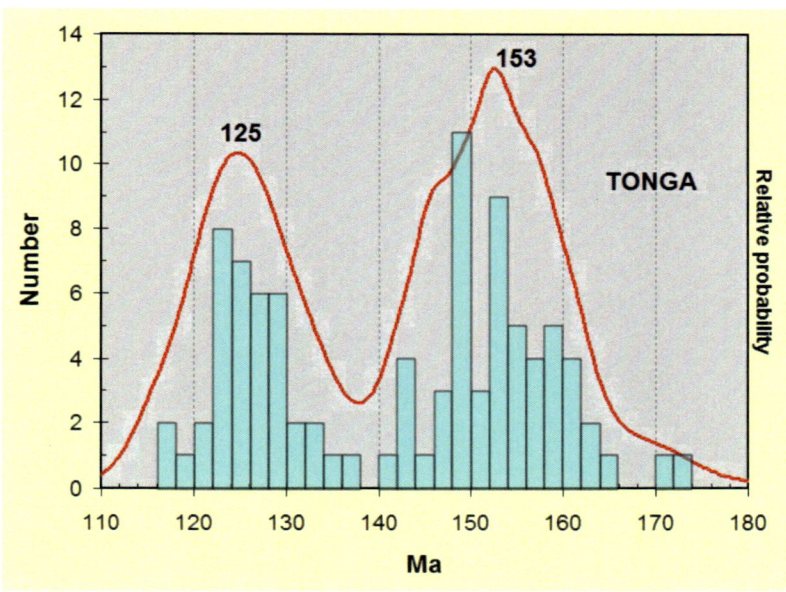

Fig. 12-9. Plot of detrital zircon ages for sandstone from the Tonga Formation. See text.

The Chiwaukum Schist is metamorphosed to the extent that, even in the lowest grade parts, age dating of the sediments by paleontology is not possible—the fossils are wiped out. So, I worked on the possibility of age analysis by way of uranium - lead measurements in detrital zircon (Chapter 3). Quartz-rich samples are the most promising because the quartz typically comes from granite, which in turn yields zircons (by laborious separation). Chiwaukum Schist is rich in quartz (Fig. 12-8), but attempts to get zircons from several samples failed. Apparently the quartz in these rocks originates from deep-sea chert (made from fossils), not continent-derived quartz sandstone eroded from granite.

I did get a good sample from the neighboring Tonga Formation. This rock unit is more-or-less continuous with Chiwaukum, separated by a pluton and a minor fault. Fig. 12-9 is a graph, that for a given age in millions of years, adds up the number of zircon grains with that age (Chapter 3). We see that the igneous sources of these zircons spread in age from about 120 to 165 Ma, with major populations at 125 and 153 Ma. This finding fits well into the regional setting interpreted for the original Chiwaukum sediments deposited in a forearc basin and sourced from erosion of the arc (Fig.4-5). Plutonic remnants of the arc at the western edge of the Quesnellia terrane are in the age range of these zircons, supporting the connection of Chiwaukum Schist to the Methow basin.

In the midst of the Chiwaukum Schist is a broad tract of gneiss, named the Nason Ridge Gneiss (Fig. 12-10). This rock is an intimate mixture of Chiwaukum Schist country rock, granitic segregations (likely formed by partial melting) and igneous injections. The injections range from outcrop-scale to mappable plutons. Such mixtures of igneous and country rock are termed migmatite, a term well applicable to the Nason Ridge Gneiss.

One interpretation of the Nason Ridge migmatite gneiss, close to melting and mixed with intrusions, is that it was once the floor of a now eroded broad overlying pluton (more later).

Fig. 12-10. Migmatite of the Nason Ridge Gneiss.

113

Swakane Gneiss

This an oddball terrane within the Cascades Core. The rock unit occurs as a bedrock panel faulted into contact with the Napeequa Schist in two places—in the southeast corner and in the central part of the region (Fig. 12-3). The rock is metamorphosed quartz and feldspar-rich sediment, probably an arkosic sand, but the initial rock type is not easily discerned because of complete recrystallization by high-grade metamorphism. The meta-sediment is locally intruded by thin granitic layers (Fig. 12-11).

Fig. 12-11. Road cut exposure of the Swakane Gneiss along U.S. 97A about 5.5 miles north of U.S. 2. Granitic sills intrude gneiss derived from sandstone. From Gatewood and Stowell, 2012. Rob Holler for scale.

Why is this rock an oddball? For one, it doesn't look like any of the neighbors—the Chelan Mtns, Nason, or Napeequa terranes. But the most distinguishing feature is the age of its zircons. A first run on obtaining zircon ages, by Jennifer Matzel and others (2004), found a broad range of ages in age-zoned grains, with the youngest at 73 Ma. The reasonable interpretation was that the original sand (protolith) from which the Swakane Gneiss was formed was younger than this age. This sedimentary age is significantly younger than found for the neighboring terranes, and implies a prolonged terrane assembly process.

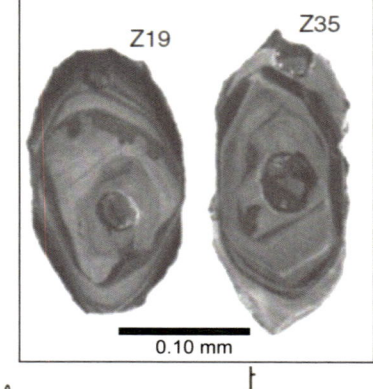

A subsequent study by Mathew Gatewood and Harold Stowell (2012), also measuring zircon ages but with the laser spot technique (Chapter 3), as

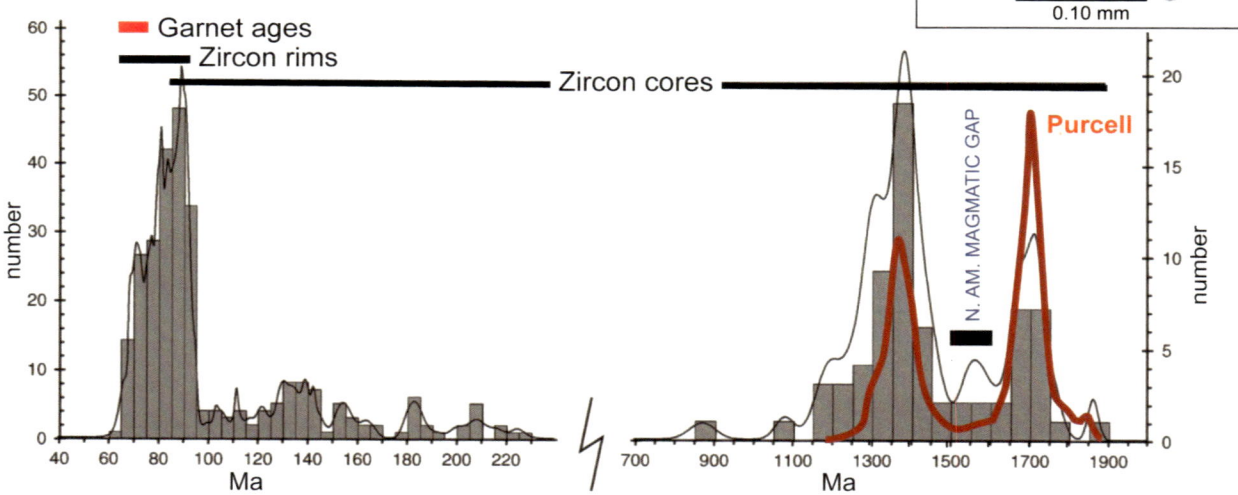

Fig. 12-12. Zircon and garnet age data. *Top Right*: Electron microscope view of Swakane gneiss zircons, with zoning revealed by cathode luminescence. Note laser ablation pits at core and rim. *Above*: Zircon and garnet ages compiled for seven samples of the Swakane Gneiss, with red overlay of zircon ages from the sandstone of the Purcell Group, southeast British Columbia. Swakane images are from Gatewood and Stowell, 2012. The Purcell zircon age data are from Gardner 2008 (Mt. Nelson). The Purcell age distribution matches closely the older ages of the Swakane Gneiss, and thus identifies a conceivable source for part of the Swakane sands.

opposed to the whole grain analysis of Matzel, also found that the zircons are zoned (Fig. 12-12); the youngest ages mostly occur as rims and could be metamorphic in age. They also dated metamorphic garnets in the Swakane Gneiss and discovered that the high-grade metamorphism, and deep burial (25 miles), ranged in age from 65-73 Ma. Their conclusion was that igneous ages of the zircons are as young as 90 Ma, and that the younger ages are metamorphic overgrowths, not igneous but still significant in dating orogeny.

A further finding of these zircon studies is a rich history recorded in zircon cores that should help us pin down the sedimentary origins of the Swakane sands. A Precambrian source spanning an age range of 1200-1800 Ma is indicated for some zircons. This age group is consistent with a source in ancient western North America, Laurentia (Fig. 4-1). Can we be more specific? Not with certainty. But, a good match is with zircons from the Purcell Group in southeast British Columbia. Zircon age peaks of about 1360 and 1700 Ma in the Swakane coincide with ages in the Purcell Group (Fig. 12-12). Such ages are also found in rocks of the southwest part of Laurentia in Arizona and southern California, so the Purcell connection is not irrefutable.

The Purcell Group consists of sedimentary rocks deposited in ancient basins in western Laurentia that developed along the juncture of Laurentia with conjoined parts of the supercontinent Rodinia (Fig. 1-7). Two age measurements in the Swakane Gneiss fall into the "North American magmatic gap" of 1490-1610 Ma, pointing to non-Laurentian input. Unfortunately, only two spot ages are not enough for certainty about foreign input, but let's consider the possible significance. Although we don't find igneous rocks of that age in North America, they do occur in Australia, and thus we can identify this foreign source by presence of zircons in the "gap".

Let's look at the whole spread of ages. Most of the ages in the Swakane zircons range from 100-200 Ma; these ages could represent igneous rocks of the ancient arc along the western edge of Quesnellia (Figs. 4-2, 4-5). There is an abundance of ages in the 70-95 Ma range. As discussed above, the youngest ages are best interpreted to be metamorphic overgrowths. Those in the 90s are likely igneous.

Where did these zircons originate? The southern Cordillera (Arizona - California) is a possibility. But more likely the source is northern Idaho or southeast British Columbia. During the Rodinia days (~1000 Ma), hills in Australia were eroding and shedding Precambrian sand into the adjacent Purcell basin in westernmost Laurentia. Our one zircon, bearing "gap" ages, possibly originated in Australia; it was eroded and deposited as sand in North America. Then, Rodinia broke up; Australia pulled away and the ancestral Pacific ocean opened. Much time passed. Purcell sediments were eroded and carried westward to the continental margin, joining detrital zircons from the Quesnellia arc, to be deposited in the Methow basin. The Cascades orogen developed, including 90 Ma plutons. The plutons intruded sedimentary country rock, picked up our detrital zircon grains, and added 90 Ma rims. Subsequently, these plutons were uplifted and eroded. Sediments from erosion of the 90 Ma plutons accumulated in a Methow related backarc basin on the east side of the Cascades arc.

The next stage in this hypothetical but possible sequence of events is the shifting of activity in the Cascades orogen to the east, into the realm of the Skagit Gneiss Complex (Fig. 12-3), at about 70 Ma. Great crustal thickening occurred, involving the Swakane sediments. Deep burial and high-grade metamorphism added yet another zone to our long-travelled zircon grains; this one metamorphic.

I would argue that the crustal thickening here occurred by magma loading. But, a case can be made against loading by plutons in that we find only small intrusions in the Swakane Gneiss (Fig. 12-11). The alternative is tectonic thickening, but there is not structural evidence of the root zone for a big thrust fault. It is a mystery unsolved.

Pluton Emplacement

Looking at the length of the whole Coast Plutonic Complex from Washington to Alaska, this mountain belt, as we have noted, is formed of two basic elements: humongous plutons, and country rock consisting mainly of accreted terranes. In terms of crustal thickening, i.e. mountain building, what is more important: piling up of country rock by folding and faulting, or emplacement of plutons (Fig. 11-2)? What can we learn about mountain building from plutons in the North Cascades?

For starters, the plutons are wonderfully exposed in alpine regions, and mountaineering geologists are happy to spend time up there. However, the emplacement mechanism of Cascades plutons is not so evident as in the Harrison Lake area, partly because igneous mineral orientations in plutons and contacts with country rock have been overprinted by metamorphism. Mostly, we cannot clearly see if the plutons are sheeted or tub-shaped; or discern strain aureoles for individual plutons distinguishable from more regional metamorphic fabrics.

Upper: **Climbing buddy Rob Coe taking a break on the ascent of Forbidden Peak, 1965. View is east toward Mt. Buckner above the Boston Glacier.** *lower:* **View west across the Boston Glacier. The sharp rock peak in the left center is Forbidden Peak; behind Forbidden, the peak with a snowy ridge above rock is Eldorado Peak; in the distance are Mt. Baker on the left and Mt. Shuksan on the right. Bedrock in the photo, out to Eldorado Peak, is the Eldorado pluton.**

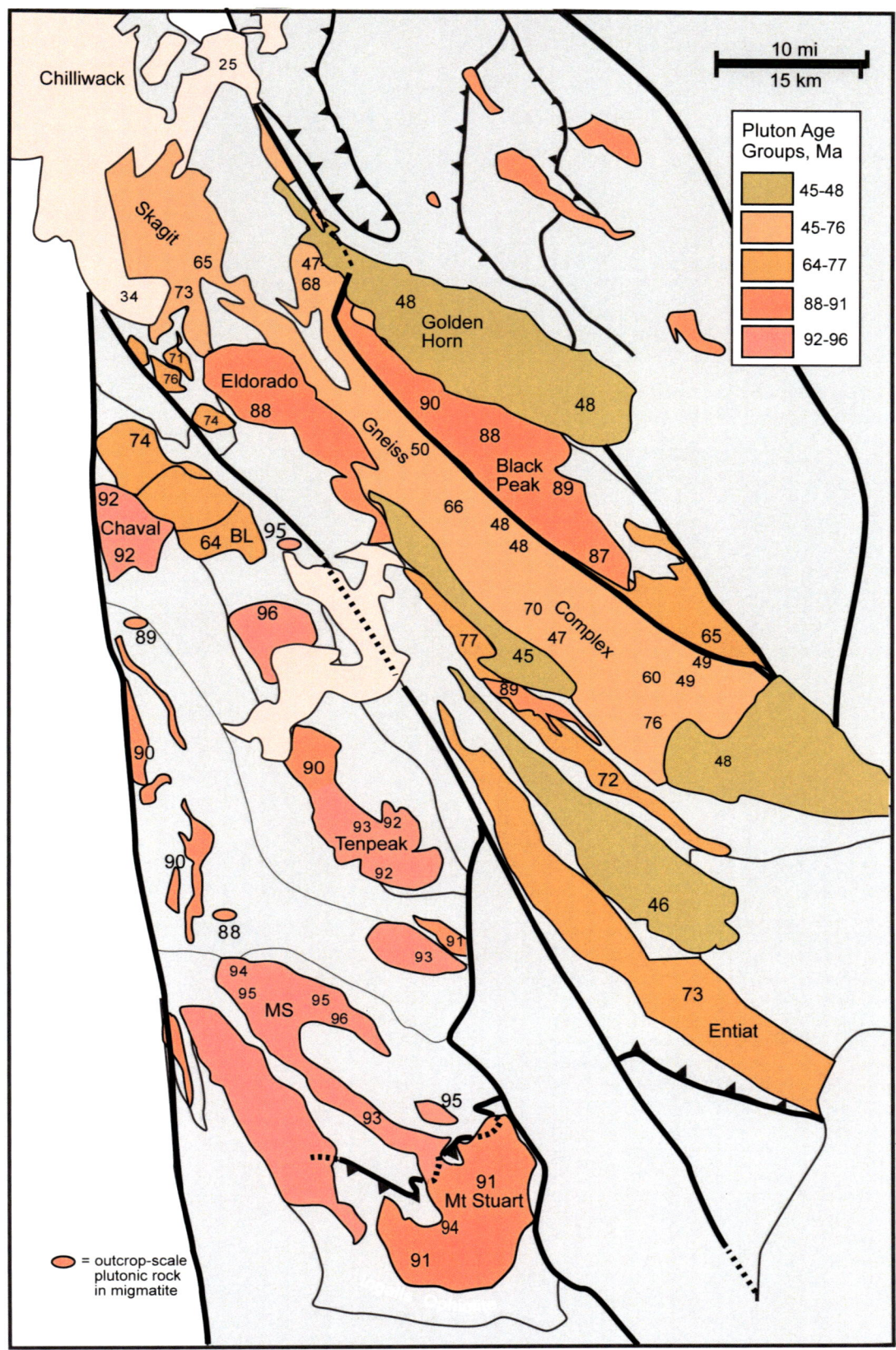

Fig. 12-13. Map of plutons, with ages in millions of years, in the Cascade Crystalline Core. Most ages are based on uranium/lead isotope ratios in zircon. Pluton age groups are defined to match those in the Harrison Lake area. BL identifies the Bench Lake pluton.

117

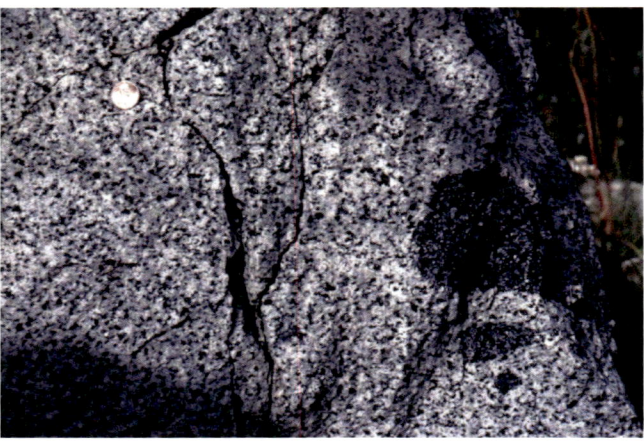

Fig. 12-14. *left*: Mt. Stuart pluton with a (rare) country rock inclusion. *right*: Also Mt. Stuart pluton, coin for scale. This sample has a dark inclusion of gabbro.

Fig. 12-15. Tenpeak pluton, with dark inclusions of hornblende-rich rock. These dark inclusions apparently formed by basalt magma injected and quenched into the much cooler quartz-diorite magma. Also, we see a late cross-cutting quartz + feldspar dike.

Fig. 12-16. Quartz diorite of the Eldorado pluton, showing an intrusive contact against a raft of country rock (a large mass engulfed by the pluton) that is apparently ribbon chert of the Napeequa Formation. The pluton is undeformed at this locality, but is an orthogneiss in other places. Photo from McShane (1992).

From zircon dating, we know that Cascades plutons occur in clumps of ages: 45-48, 64-77, 88-91, 92-96 Ma (Fig. 12-13). Plutons of these age groups are spatially intermixed, but in a rough sense they group in belts parallel to the length of the orogen, that decrease in age from west to east; this means the source of magma in the arc migrated eastward with time. The Cascades plutons have counterparts in the Harrison Lake area. Younger plutons (64-77 Ma) are not well documented in British Columbia; they are possibly there but not identified due to sparse zircon geochronology. Younger plutons in the Cascades, 3-35 Ma, are scattered and relatively minor; together with active volcanoes they are an expression of the ongoing magmatic arc system.

A recent study of the Golden Horn pluton is of special interest to seekers of the mystery of pluton origins in the Cascades (Figs. 12-17, 12-18, 12-19). It was long known from the pioneering work of Peter Misch in the 1950s that the pluton is made up of several varieties of granite, distinguishable based on textures and mineralogy. The pluton is roughly dated at 48 Ma.

Mike Eddy of M.I.T. set out to very precisely map and zircon-date these different parts of the pluton. It's a large pluton, accessed in terrain with considerable relief, allowing a 3-D analysis. The rock is younger than the regional metamorphism, thus original

Fig. 12-17. Panorama of peaks carved out of the Golden Horn pluton, viewed from the "Washington Pass Overlook". Peaks on the right skyline are Liberty Bell, the highest peak, and Early Winter spires to the left of Liberty Bell. A switchback in highway 20 cuts into the pluton.

pluton features are not wiped out. I attended a lecture Mike gave on this project. He explained that, as he hiked up the sides of ridges and mountains, the rock changed from one type of granite to another. And as he worked along from one traverse to the next, climbing up the topography, the contacts between two granite types maintained a fairly constant elevation. He recognized six types of granite defining approximately horizontal sheets (Fig.12-18). The sheets are locally displaced by steep faults. The total pluton thickness exposed and measured is 7-8 km (~5 mi); the floor of the pluton is not exposed, lying at an unknown depth. Zircon ages of the individual granite sheets are accurate to +/- 30 thousand years (remarkably precise) and indicate that each sheet has a separate and distinguishable age. The sheets were injected separately; that is, they were not layers in a single large magma chamber. The age of sheets decreases upward, showing growth of the pluton by additions at the top. Total length of time for emplacement of the six sheets was about 750,000 years.

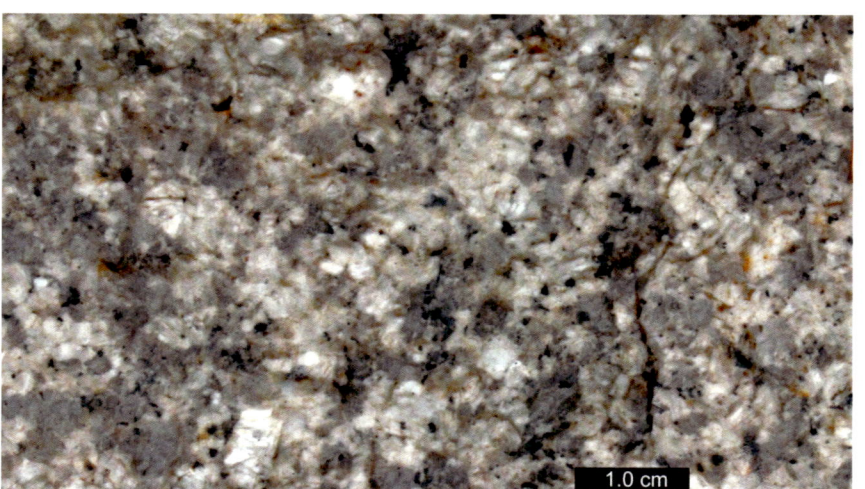

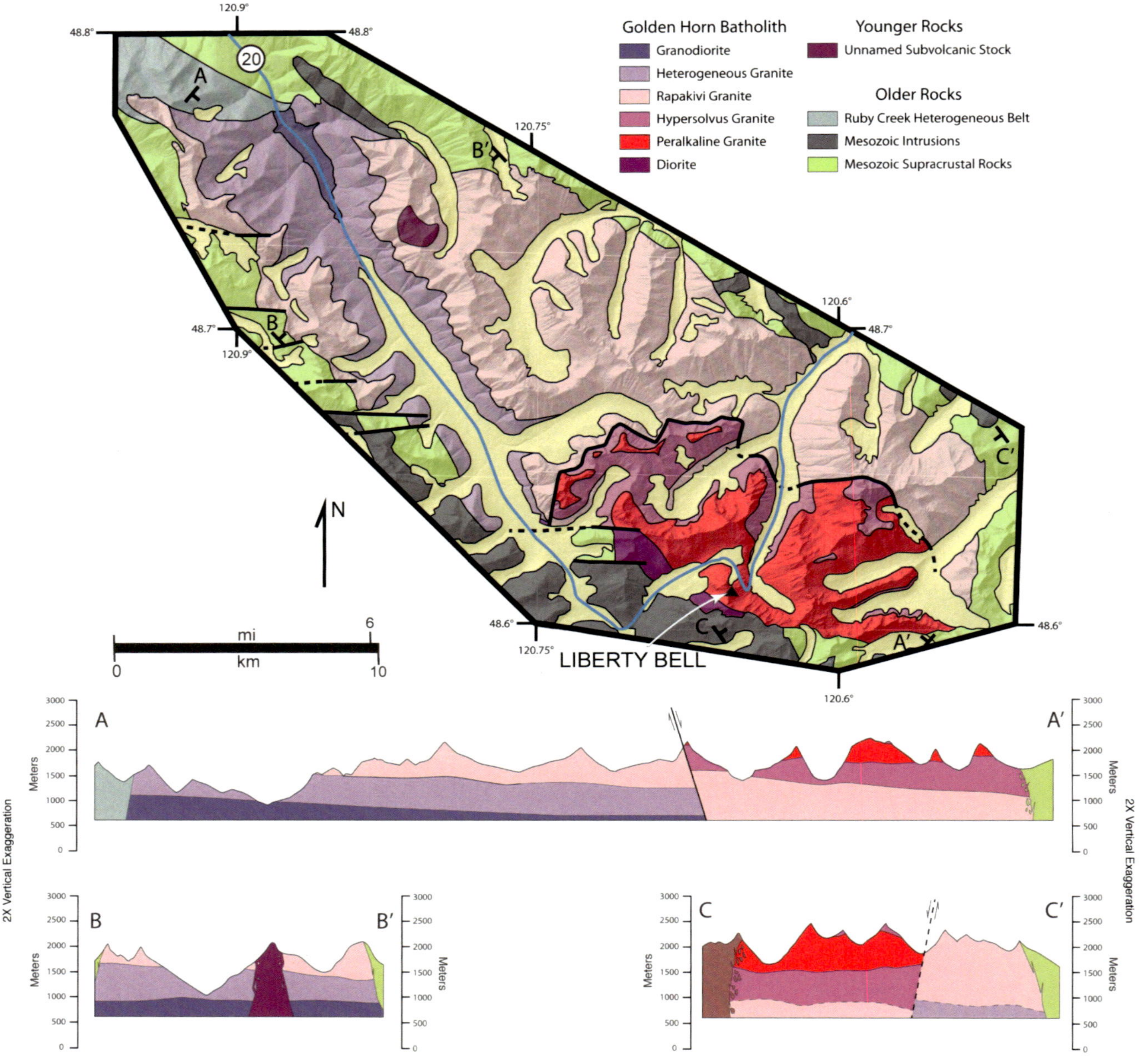
Golden Horn Batholith
Granodiorite
Heterogeneous Granite
Rapakivi Granite
Hypersolvus Granite
Peralkaline Granite
Diorite
Younger Rocks
Unnamed Subvolcanic Stock
Older Rocks
Ruby Creek Heterogeneous Belt
Mesozoic Intrusions
Mesozoic Supracrustal Rocks
20
N
mi
6
km
0
10
LIBERTY BELL
120.9°
48.8°
120.75°
120.6°
48.7°
48.6°
A
A′
B
B′
C
C′
Meters
2X Vertical Exaggeration
3000
2500
2000
1500
1000
500
0

Fig. 12-18. *above*: Map and cross sections of the Golden Horn pluton, from the publication of Eddy et al. 2016. Illustrated is the sheet structure of the pluton.

1.0 cm

Fig. 12-19. *left*: Granite from the roadside near the base of Liberty Bell peak. This rock as mapped by Eddy is a "peralkaline granite", bearing Na amphibole—dark specks in the photo. Other minerals are grey quartz and white feldspar.

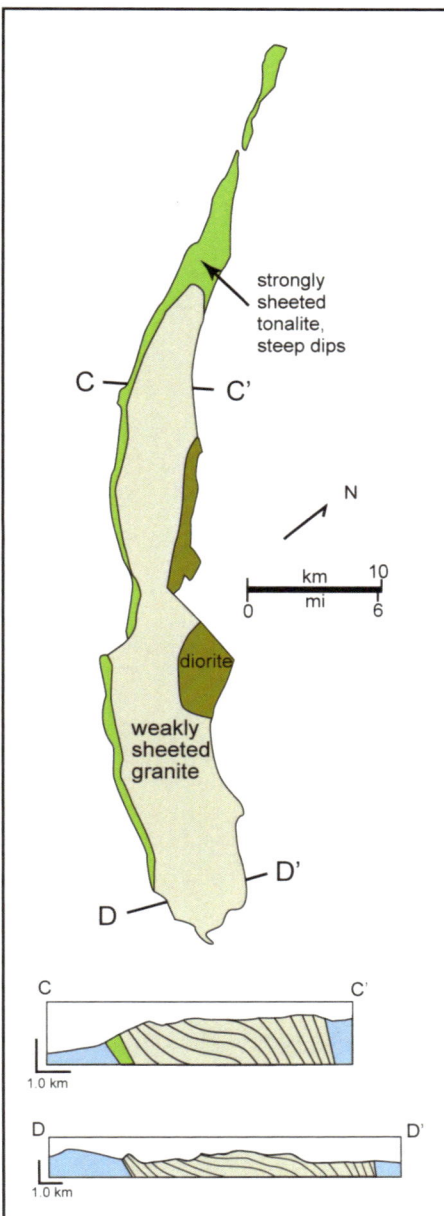

Fig. 12-20. Map and cross sections of the Entiat pluton, adapted from Miller and Paterson, 2001. Structure shown within the pluton is defined both by sheeting and magmatic foliation. In the cross sections, the trace of sheets is projected to depth from surface

The horizontal sheet layering of the Golden Horn batholith finds a good match in the much older Breakenridge pluton of the Harrison Lake area (Fig. 11-6).

Two other plutons in the Crystalline Core have preserved magmatic fabrics and a discernable sheeted structure. These are the Cardinal Peak and Entiat plutons, described by Bob Miller and Scott Paterson (2001). These plutons are a component of the broad and elongate assemblage of 45-76 Ma plutons and Napeequa Schist country rock that includes the "Skagit Gneiss Complex" of Fig. 12-13.

Both plutons show a border region of steeply dipping fabric and sheet structure. But also both plutons exhibit a broad inner zone of relatively low-dip sheets and foliation (Fig. 12-20). From this, Miller and Paterson interpret the pluton emplacement mechanism to be dominated by assembly of vertical sheets, and by diapirism. The low-dip zones they attribute to tectonic reorientation of originally vertical sheets. There is evidence of tectonism affecting the plutons—some of the fabric formed after solidification of the magma, and country rock is deformed.

The low-dip zones pretty much dominate the pluton fabric. In this light, and considering the structure and emplacement of the Golden Horn pluton, an alternative explanation for the Cardinal Peak and Entiat plutons is that they formed originally as a stack of horizontal sheets, with subsequent regional tectonics steepening the marginal zones.

Miller and Paterson suggest that these two plutons could have served as vertical sheet-like conduits moving magma from a deep source upward into large magma chambers, as apparent for plutons of the Harrison Lake area. They further propose that the steeply sheeted zones of the Entiat and Cardinal Peak plutons are relicts of roots to upward welling diapirs (Fig. 11-3). As an alternative to the diapir explanation, we could consider that the zones of near horizontal sheets are the floors of broad horizontally sheeted plutons, now mostly eroded away.

Finally, in the review of plutons in the Cascade Core, we'll look at the northern part of the Skagit Gneiss Complex (Figs. 12-21, 12-22), traversed by Highway 20 and thus quite accessible to geologists and anyone else with an eye for rocks. Debate continues as to the cause of the huge vertical crustal displacements of the Skagit Complex, 20 miles down and then up, in a time span bracketed by ages from 88-60 Ma. At the southwest edge of the Skagit Complex is the Eldorado pluton (Fig. 12-13), dated by zircons at 88 Ma, and initially intruded at a relatively shallow level (< 5 miles) indicated by relict andalusite and zoned garnets (Figs. 12-26 & 12-28, respectively).

Moving north from the Eldorado pluton into the nearby Skagit Complex, we find gneissic metamorphosed plutons (Fig. 12-21A) in the age range of 75-65 Ma, and unmetamorphosed plutons with an age of about 45 Ma. Napeequa Schist country rock, strongly metamorphosed and termed paragneiss, is mixed with lenses and layers of the gneissic igneous rocks, and is cross-cut by the post-

metamorphic plutons (Fig. 12-21 B&C). Migmatite is an appropriate term for this mixture of plutonic rocks and metamorphic country rock. Measured peak temperature, based on mineral compositions, is in the range of 700 °C, on the verge of melting. Pressures are in the 9-10 kb range, equating to 20 miles burial. We'll come back to origins of this deeply buried complex in the next section as we look at the metamorphic effects.

Fig. 12-21. Metamorphic rocks of the Skagit Gneiss Complex. A : Metamorphosed pluton, an "orthogneiss". Strong NW-SE lineation indicates orogen-parallel movement. Jim Talbot photo. B & C: Skagit migmatite. Gneissic country rock (Napeequa Schist ?) is mixed with granitic intrusions, as Jim Miltimore points out (~1971).

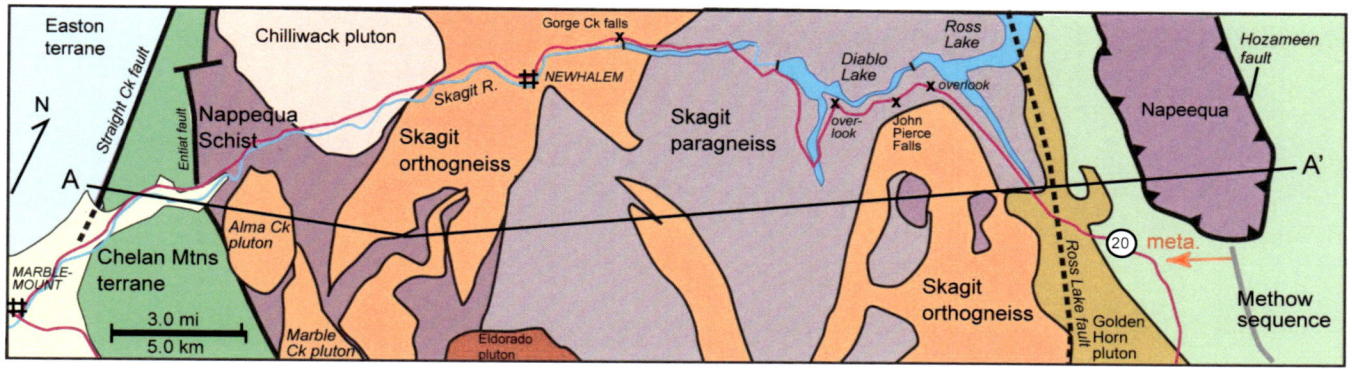

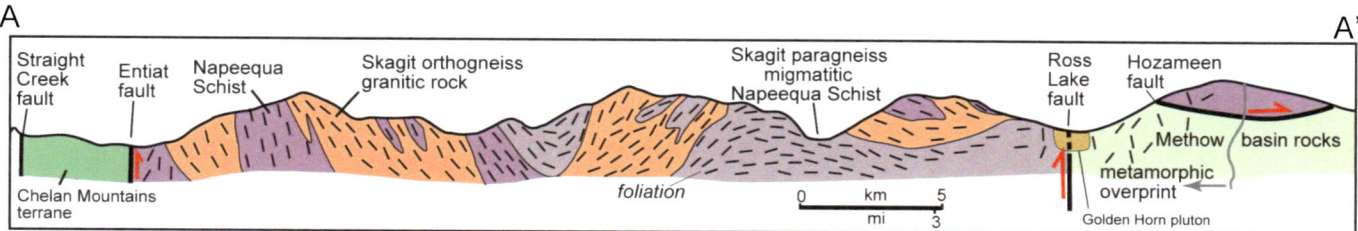

Fig. 12-22. *above*: **Map and cross section of the very high-grade metamorphic complex of the Skagit Gneiss Complex, adapted from Tabor et al., 2003, and Haugerud and Tabor, 2012. The originally surficial rocks of the Napeequa terrane (sediments) were deeply buried, intruded by plutons, and subsequently uplifted some 20 miles, all in the time frame of 88-60 Ma. The cause of this burial is a burning question.**

Fig. 12-23. Professor Peter Misch of the University of Washington on a 1967 field trip into the Skagit Gneiss Complex. Misch left Nazi Germany in the 1930s and came to Washington by way of geologic research in the Alps and Himalayas, and in academic positions in China. In the 1950s and 60s, he and students at UW delineated the basic structural framework of the North Cascades.

View west from the Diablo Lake overlook along Highway 20. Bedrock of the landscape is the Skagit Gneiss Complex. The lake is created by damming of the Skagit River, and is managed for power by Seattle City Light. The vivid blue-green lake owes its color to glacial meltwater draining from high peaks to the south.

Metamorphism

Country rock and most plutons in the Cascades Crystalline Core have undergone (enjoyed?) some degree of recrystallization and textural overprint owing to heat, confining pressure, and strain. In the sense that this metamorphic effect can obscure primary pluton igneous fabric, it is unwelcome. But, nonetheless, the overprint mineralogy and fabrics of the metamorphism reveal an intriguing story of the mechanisms of crustal thickening and mountain building. Not surprisingly, much of the findings here are of the same sort as in the Harrison Lake area.

Country rock near plutons commonly bears pseudomorphed andalusite (Figs. 12-24, 12-25), a mineral that in its original form signals a relatively shallow depth of metamorphism < 5 miles (Fig. 3-7, Chapter 3). We can surmise that the andalusite formed by contact metamorphism caused by the nearby pluton.

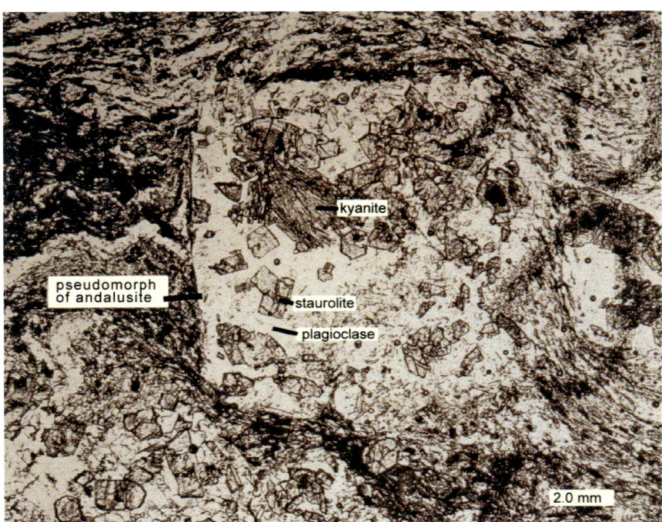

Fig. 12-24. Andalusite replaced by a higher pressure mineral assemblage, including kyanite. Chiwaukum Schist at the northern edge of the Mt. Stuart pluton, same for the sample below.

Fig. 12-25. Chiwaukum schist showing large white and brown mineral aggregates of kyanite, feldspar and staurolite replacing andalusite. Boulder from Nason Creek near Highway 2.

What replaces the andalusite? More metamorphic minerals, indicating a deeper level of metamorphism, as much as 20 miles below ground surface. These replacement minerals include kyanite, staurolite, feldspar, and others; and they occur throughout the metamorphic rock, not just within the original andalusite crystal. This mineralogy and replacement texture is the same as that seen in the Settler Schist of the Harrison Lake area.

Thus, over much of the country rock region, we find a two-step process in the metamorphism: an early shallow event and a later deep event. The early event is reasonably attributed to plutonism, but the deep metamorphism is not so simply understood, and has evoked various interpretations. One is that the high-pressure metamorphism was caused by structural thickening of the earth's crust by contraction across the orogen in the form of thrust faulting or tight folding (Fig. 11-2A). Certainly in some other mountain belts this process was operative (Fig. 2-2).

An alternative explanation is that the crustal thickening was caused by plutons rising and displacing country rock downward (Fig. 11-2B).

Pluton shapes, structures, and relation to country rock in the Harrison Lake area are more consistent with plutonic rather than structural loading (Chapter 11). But, the case is not so clear for the Cascades, where most workers in recent years have embraced the structural model. Let's review the evidence and reasoning.

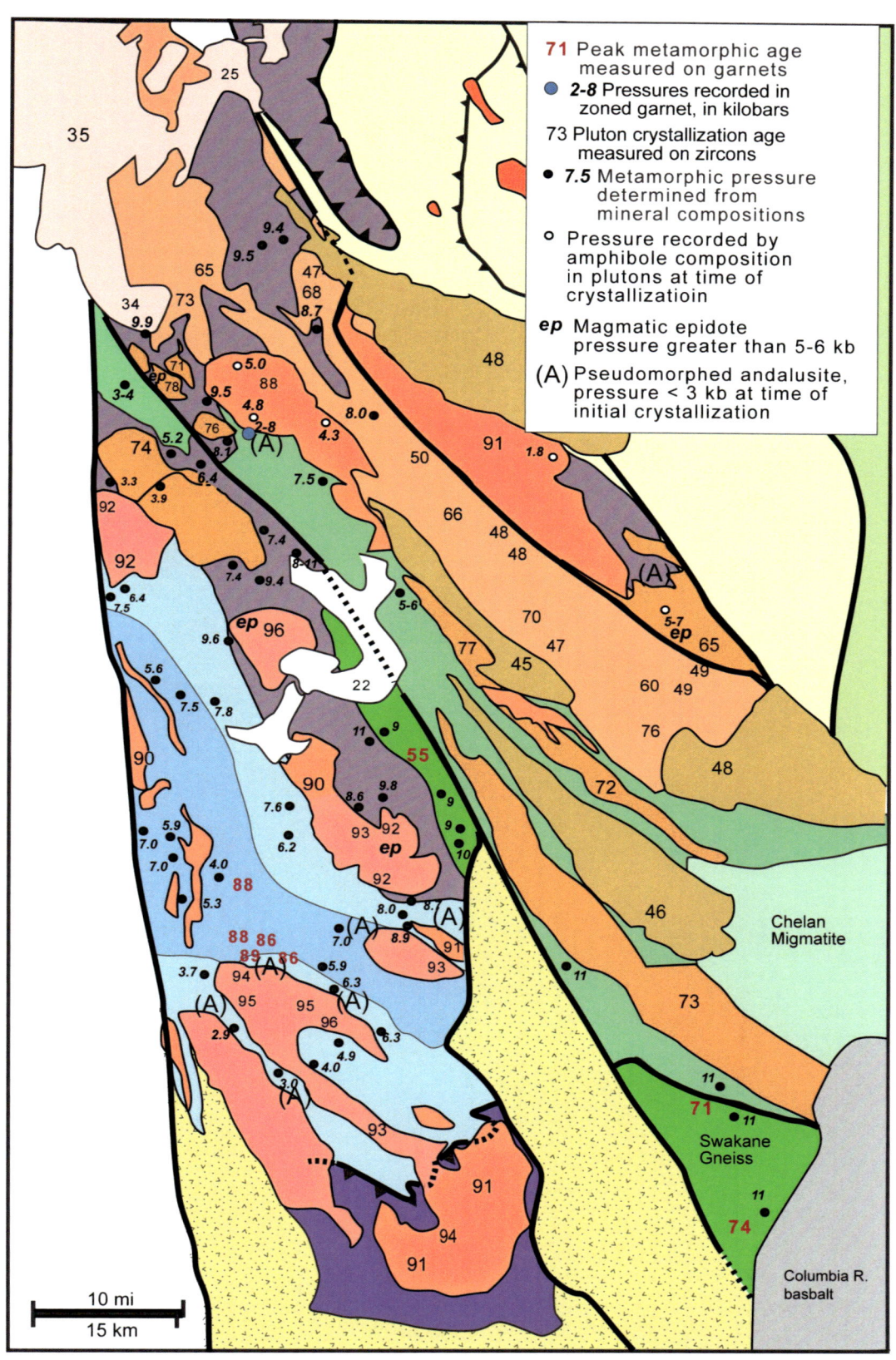

Fig. 12-26 Map of measured metamorphic pressures, metamorphic ages, igneous pressure of crystallization, and pseudomorphed andalusite indicating early metamorphic pressure < 3 kb. 1.0 kb = 2.3 miles burial.

125

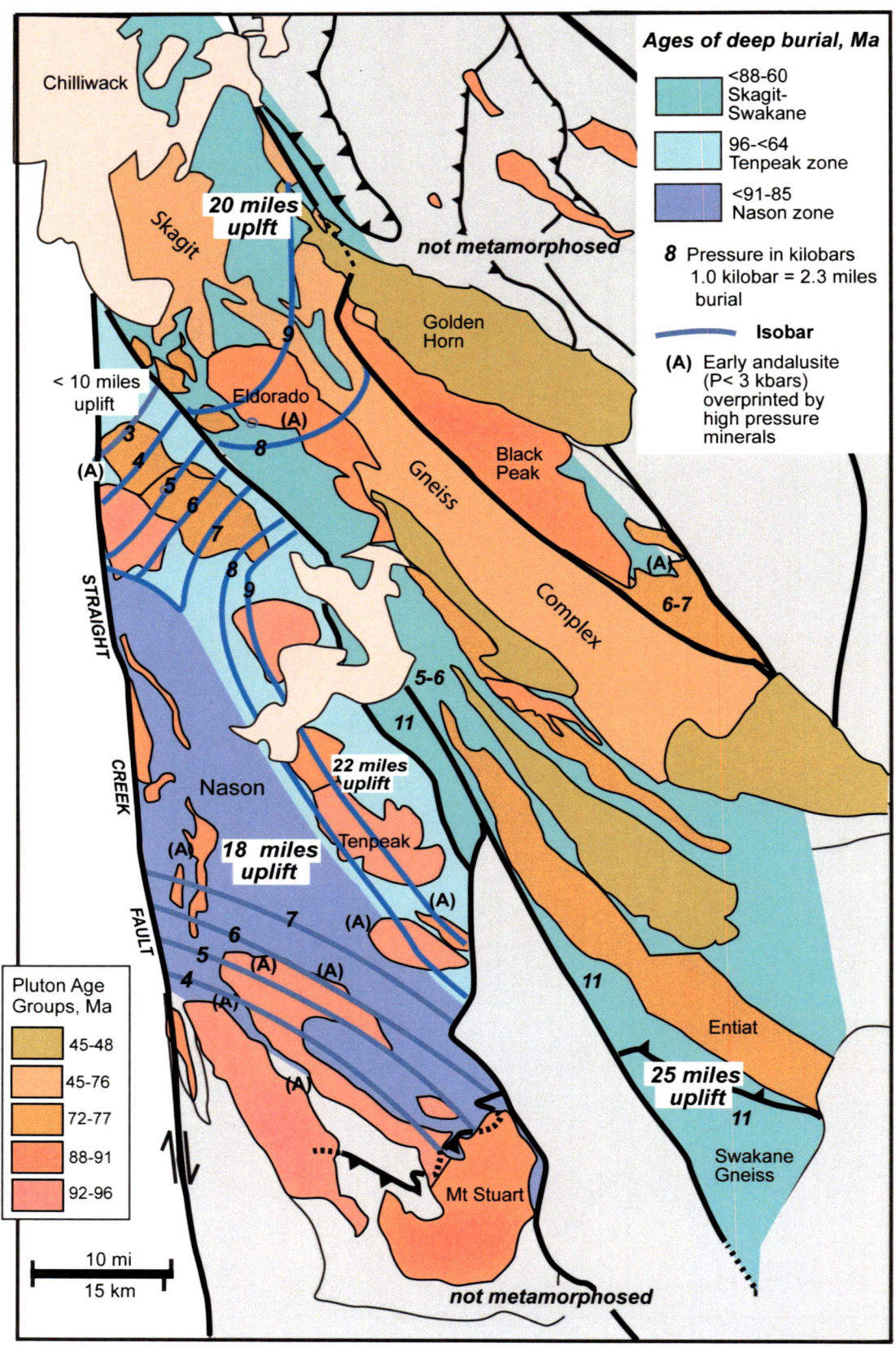

Fig. 12-27 Three zones of deep burial and uplift in the Cascades Crystalline Core. Zones are distinguished by differing ages of subsidence and uplift.

An impressive amount of data derived from sophisticated laboratory procedures has become available for the Cascades in the past 20 years (Chapter 3, reviewed briefly here). We know pluton ages from uranium-lead isotope analyses of igneous zircon in the plutons (amazingly precise in the case of the Golden Horn pluton). The age of metamorphism is measured in garnets by isotope ratios in the samarium-neodymium chemical system, a relatively new procedure not so widely applied yet. Potassium-argon dating of mica and amphibole gives metamorphic ages for low-grade rocks, or a cooling age for high-grade rocks. Metamorphic pressure and temperature are measured by kinds of minerals in the rock and by mineral compositions (Fig. 3-11). Pressure during pluton crystallization is measured by the aluminum content of hornblende, and also by the presence of epidote, which occurs as an igneous mineral only at depth greater than about 12 miles.

Values for ages and pressures measured for country rock and plutons are plotted in Figs. 12-26. This is more detail than the non-geologist would like to see, but it is an important data base for speculation about what part of the earth's crust went up or down and when. This mass of data is interpreted in Fig. 12-27 in terms of a "big picture" of the amounts, regions, and timing of crustal loading and unloading.

Three regions characterized by differing times of up and down crustal mobility appear:
1) A zone central to the orogen, including the Tenpeak pluton, that was loaded by 95 Ma indicated by epidote-bearing plutons. The uplift age is not well known, but predates a 64 Ma hornblende potassium-argon age of the nearby Bench Lake pluton (Fig. 12-13). The amount of burial (and uplift) ranges from 22 miles in the Tenpeak area, to less than 10 miles in the vicinity of the Cascade River Valley (Fig. 12-30).

2) Flanking the Tenpeak zone on the southwest is the Nason zone, centered on the Nason Ridge Gneiss and including the Chiwaukum Schist (Figs. 12-3, 12-27). Loading occurred after the 94-96 Ma parts of the Mt. Stuart pluton, the andalusite-bearing aureole of which is overprinted by kyanite (Figs. 12-24, 12-25). Peak metamorphic garnet ages are 88-89 Ma. A zircon age of migmatite in the Nason Ridge Gneiss is 89 Ma. Potassium-argon ages indicating uplift are as young as 85 Ma. Here, burial and uplift of about 18 miles occurred between 94 and 85 Ma.

3) The largest loaded zone, both in breadth and length, is the Skagit Gneiss Complex on the east flank of the orogen. Loading occurred after the 88 Ma Eldorado pluton, which has pseudomorphed andalusite and strongly zoned garnet in its aureole (Fig. 12-28). Very high pressures of up to 11 kilobars are found in this belt (Fig. 12-26), indicating uplift of as much as 25 miles. The timing of uplift is marked by ages of lower pressure minerals sillimanite and cordierite dated at 60 Ma, and by potassium-argon biotite ages of 45 Ma.

To a large degree, plutonism is found to be associated with the loading events (Fig. 11-27), both in location of

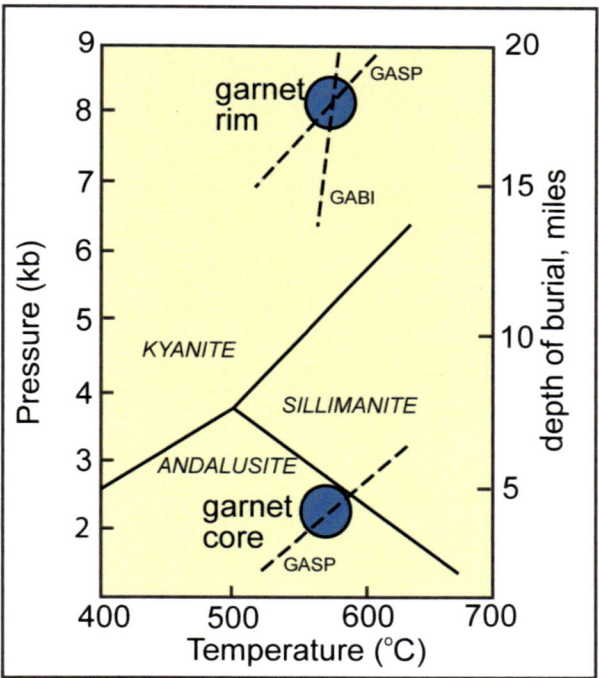

Fig. 12-28. Zoned garnet, associated with pseudomorphed andalusite in the aureole of the Eldorado pluton (88 Ma). A fifteen-mile thick load is added to the region of this sample sometime after 88 Ma. The location is marked "2-8" on Figs. 12-26, and 12-30. This great find, in very rough country, was made by Dan McShane as part of his M.S. thesis.

plutons, and in their ages. The central zone is associated with the Chaval, Sulphur Mountain, Tenpeak and Wenatchee Ridge plutons. Also of this age is the older part of the Mt. Stuart pluton, but this body lies outside the central loaded zone.

The Nason zone is intruded by the Sloan Creek plutons, and contains small-scale plutons of similar age in the central migmatitic portion. But, no large plutons occur in this zone.

The Skagit Gneiss Complex exhibits an impressive concentration of plutons in an age-range of 65-75

Fig. 12-29. *right*: **Terranes and plutons of the Cascade River area. MMQD = Marblemount Meta Quartz Diorite, and CRS = Cascade River Schist—both units comprising the Chelan Mtns. terrane. NP = Napeequa Schist terrane. Stretching and mineral lineations, all with shallow plunges, are shown.**

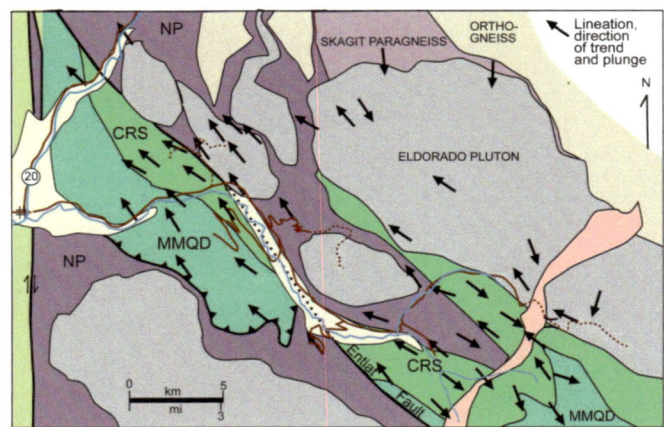

Fig. 12-30. *below*: **Barrovian metamorphic zones (Fig. 11-24), metamorphic and igneous pressures, muscovite metamorphic age, and pluton ages in the Cascade River area.**

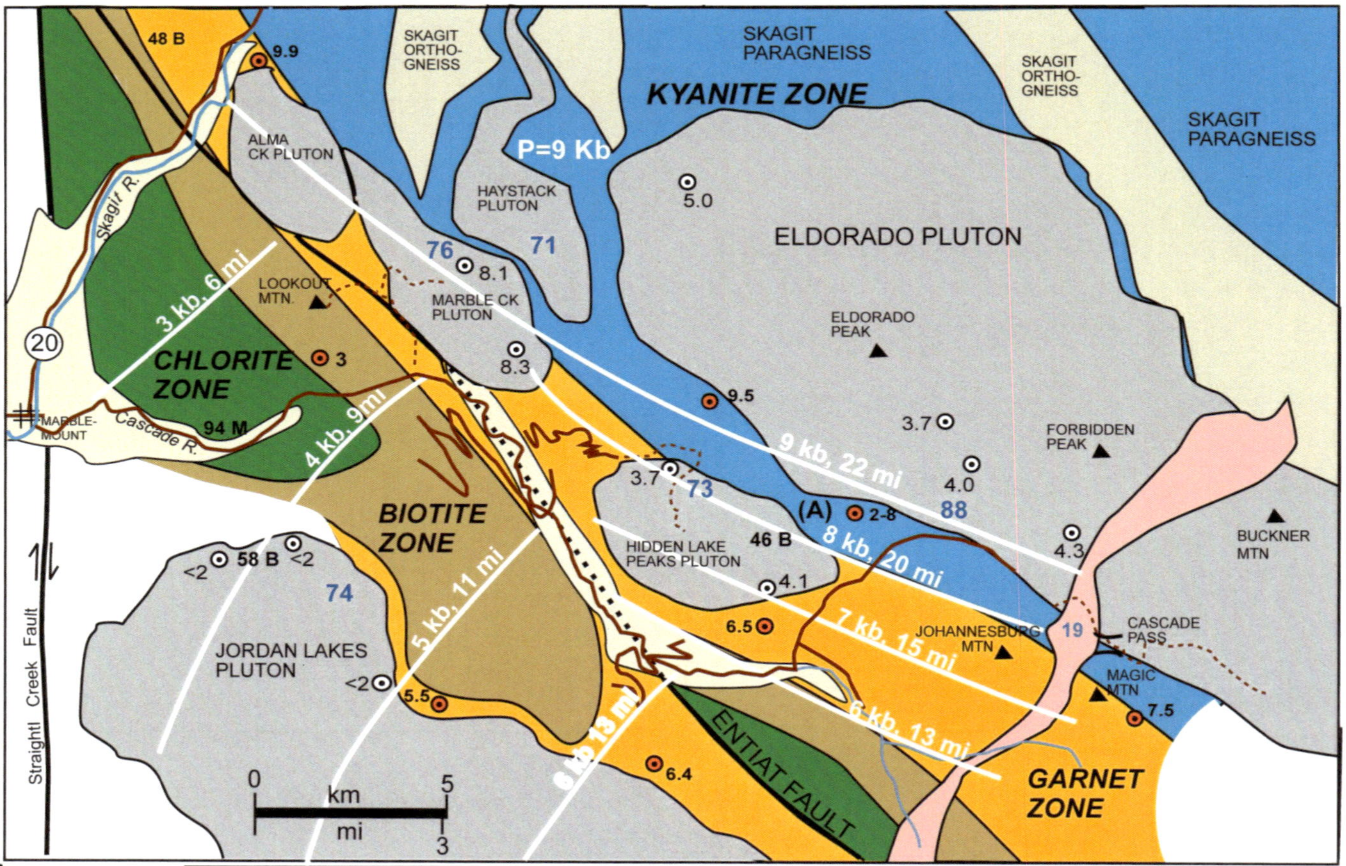

⊙ 3.7 Pressure of crystalization of pluton measured by Al in hornblende, kilobars

◉ 6.4 Pressure of metamorphism measured by mineral compositions, kilobars

94 M Crystallization age in Ma of muscovite mica in low-grade metamorphic rock, measured by K/Ar

46 B Cooling age in Ma of biotite in igneous rock, measured by K/Ar

74 Zircon age, Ma, of igneous rock

128

Ma, very close to constraints on the timing of high pressure metamorphism: older than 60 Ma and younger than 88 Ma. Mixed in with these plutons is a younger batch of 45-48 Ma plutons that were intruded after the loading and uplifting of the regional metamorphic event.

We have defined three regional zones in the Cascades in each of which metamorphism and plutonism are seemingly associated, and between which are differences of timing. Our quest is to understand the overall tectonic processes building these roots of the orogen, and we'll come back to look at these regional zones in that light farther along in this chapter.

Details in the Cascade River area help with the understanding of regional tectonics (Figs. 12-29, 12-30). This region straddles the transition from the 65-75 Ma Skagit metamorphic-plutonic zone to the 91-95 Ma Tenpeak plutonic zone. So, what do we learn from the Cascade River area?

Barrovian metamorphic zones overprint terrane boundaries, so we know deep crustal loading developed after terrane assembly.

The Skagit zone is truncated by the Entiat Fault, especially notable in the Lookout Mountain area, where the pressure jump is from 8-9 kb down to 3 kb—equating to an uplift of ~15 miles of the Skagit side relative to the low-grade side of the fault. It is apparent that the Entiat fault is younger than Skagit metamorphism, dated at about 65 Ma.

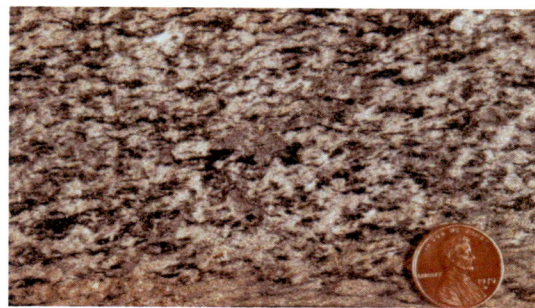

Fig. 12-31. Igneous fabric overprinted by metamorphic strain fabric in the Eldorado pluton.

The Eldorado pluton, a broad mass of granitic rock, is entirely within the Skagit loaded zone, and was intruded prior to loading evidenced by zoned garnet in the contact aureole at its south flank (Fig. 12-28). What marks does the Eldorado pluton bear of the loading? According to Dan McShane (thesis), the internal parts of the pluton, about 30 percent of the body, show little or no strain effect—the magmatic textures are preserved (e.g. Fig. 12-16). Strain fabric caused by deformation is seen in the more outer parts of the pluton body (Fig. 12-31). This fabric exhibits northwest trending mineral lineations (Fig. 12-29), apparently pointing to an orogen-parallel shear displacement. Whatever the loading process was, much of the igneous fabric of the Eldorado pluton was unaffected.

Metamorphic lineation, in the form of stretching of relict sedimentary grains and alignment of metamorphic minerals, is pervasive across the transition zone from Skagit zone into the Tenpeak zone

(Figs. 12-29, 12-32). The lineations are mostly shallow plunging, <25°, trending northwest-southeast—parallel to the length of the orogen. The lineations could be the result of simple extrusion of the orogen along its length, or they could have formed by shear as in the Easton Metamorphic Suite (Fig. 10-19). I favored the shear model because in places the deformed grains show an asymmetry indicating a sense of

Fig. 12-32. Stretching lineations in conglomerate of the Cascade River Schist.

rotation due to one side moving relative to the other. The sense of shear is mixed, but mostly dextral. Our measurements found 15 dextral, 3 sinistral over the region. Notably, the lineations support shear parallel to the length of the orogen, not contractional shear across the orogen.

Cause of Burial and High-Pressure Metamorphism

It would seem that the stars are aligned for interpreting burial and high pressure metamorphism of the Cascade Core area to be caused by orogen-normal contraction (northeast-southwest), as originally suggested by Peter Misch (1966). Deep burial by thrusting in collisional zones is well documented in other mountain belts, e.g. the Alps (Fig. 2-2). A great stack of nappes lies near the Cascades Core, in the San Juan Islands - northwest Cascades thrust system. The age of thrusting of these nappes is synchronous with Cascade Core metamorphism— thus, it's a good bet these thrust sheets rode over and buried the Core. As the cause of contraction, we have Wrangellia, a large terrane roaming about off-shore of the Pacific Northwest on the Farallon Plate and potentially poised for collision with western North America at the time of Cascade Core metamorphism.

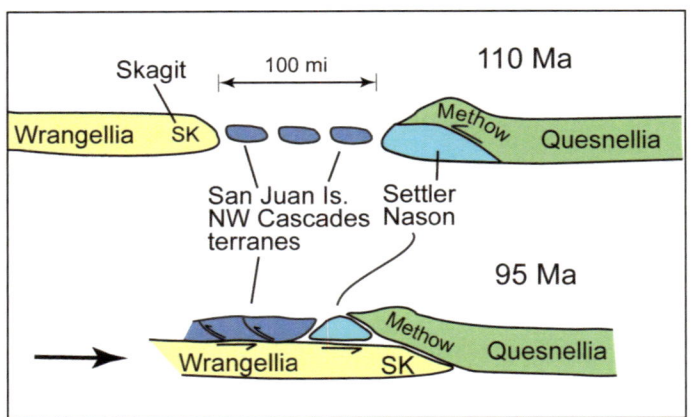

Fig. 12-33. McGroder (1991) model for burial of the Skagit Gneiss, view north. Wrangellia terrane closes against the edge of North America. First, Methow terrane is thrust over its outboard edge creating Settler Schist and Nason metamorphic belts. Further collision thrusts terranes of the San Juan Islands and northwest Cascades over the Skagit domain, causing deep burial and metamorphism. In total, some 300 miles of orogen-normal contraction occurred.

Mike McGroder (1991) embraced the Misch model (Fig, 12-33), invoking some 300 miles of closure and contraction between Wrangellia and the continental margin. The nappes were thrust out of a "root zone" (Fig. 11-2A) between the Skagit complex and the edge of Quesnellia, riding over and loading the Skagit area.

But, there are problems with this vision. As outlined in Chapter 5, p. 46, all evidence points to the nappes of the San Juan Islands - northwest Cascades thrust system having moved northwest onto the southeast edge of the CPC, not across the CPC orogen: there is no evidence of a root zone, and no mark in the nappes of CPC plutonism, nor of high temperature - low pressure metamorphism that would have obliterated aragonite.

A different loading-by-contraction model is presented by Brian Wernicke and Stephen Getty (1997). These authors, noting a lack of thrust faults exposed along the flanks of the Skagit Gneiss, developed a tectonic scenario for burial by localized "downwelling" internal to the gneiss complex (Fig. 12-34). Again, contraction by collision of Wrangellia is the driving force. This process, including the uplift, is envisaged to have involved considerable ductile flow as the rocks are squeezed, pushed in a bulge downward, and spread laterally at depth. Subsequently, as contraction stops, the bulge flows back to the surface and is unroofed by erosion.

This concept is not much supported by observed structures in the Skagit domain. Structural geologists working in the Skagit Gneiss area do see large-scale open folding (Fig. 12-22), but large high-strain structures, faults and folds, are not mapped.

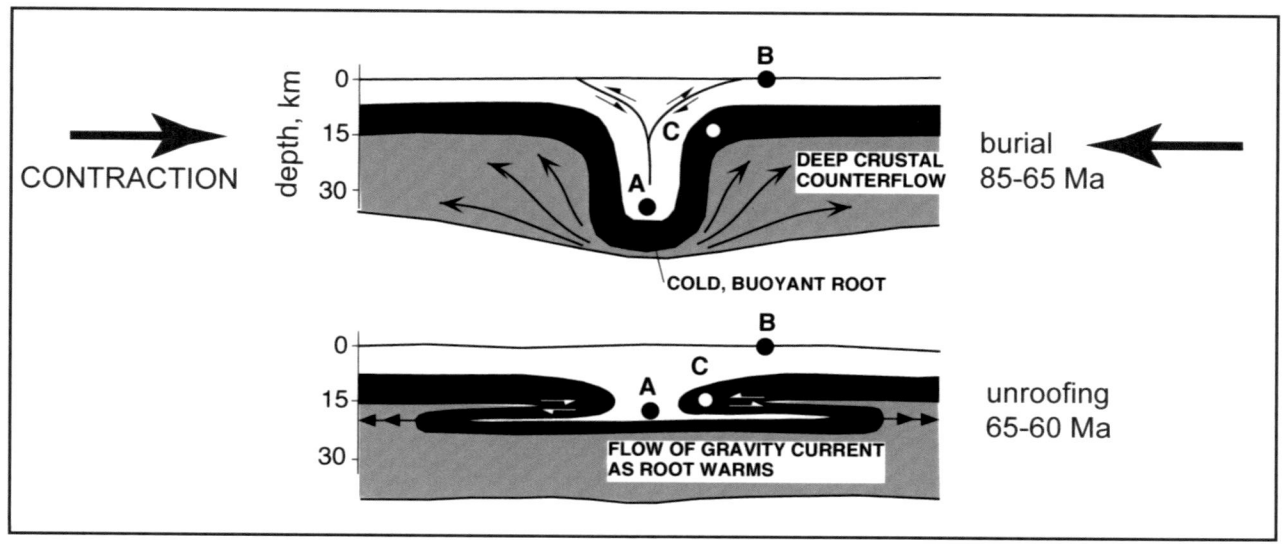

Fig. 12-34. Tectonic model of Wernicke and Getty (1997) for burial and unroofing of the Skagit Gneiss. This explanation is based on interpretation that in mid-Cretaceous times accreted terranes of the Pacific Northwest experienced, on a regional scale, an event of great orogen-normal contraction, causing much crustal shortening and burial across the Skagit Gneiss area. The mechanism invoked is "localized downwelling", leading to the high pressures of metamorphism. A, B, and C mark the paths of once near-surface rocks. Subsequent isostatic adjustments after contraction floated the deeply buried rocks upward. Further uplift and unroofing occurred in later times. In this model the rocks achieved high pressure and temperature minerals, and also high strain.

The Eldorado pluton subsided en masse to the depths of burial; it is not broken-up or smeared-out. The pluton is foliated in places, but over large areas retains an igneous fabric (Fig. 12-16); this is not what one would expect from a tightly contractional "downwelling" ductile flow process.

In addition, arrival of Wrangellia in our neighborhood is widely regarded as occurring by about 100 Ma, long before Skagit Gneiss metamorphism (>60, <88 Ma), and as a strike-slip event, not collisional (Chapters 4, 5).

What about older parts of the Cascade Core, the Nason terrane? Large-scale orogen-normal thrust faults are not found within or flanking the Nason terrane, either in country rock or plutons. However, detailed analyses of outcrop-scale rock fabrics in the Nason terrane by Bob Miller and Scott Paterson (and colleagues, 2006) finds super-position of multiple generations of tight folds and reverse faults. Unravelling the structural history related to high pressure metamorphism points to orogen-normal contraction. Conceivably, crustal thickening by orogen-normal compression could have happened, not by a giant thrust sheet, but by bulk shortening (Fig. 11-2A), i.e. like a block of butter squeezed between the jaws of a vise.

Can the Harrison Lake geology help us understand the Cascades? Let's look at the regional geology before movement on the Straight Creek - Fraser River Fault (Fig. 12-35). The Harrison Lake rocks were just next door and along strike to the Nason Ridge Gneiss and Chiwaukum Schist, inviting us to consider extension of Harrison Lake geology into the Cascades. At pressures of 7-8 kb, and metamorphic ages of 88-90 Ma, the Nason terrane is a candidate for burial by the 84-92 Ma plutons of the Scuzzy-Urquhart assemblage. In the Nason terrane, we could be looking at the substrate below large tub-shaped plutons now mostly eroded away. Remnants of these large bodies would be the 88-91 Ma small plutons and igneous parts of migmatite in the Nason terrane.

Finally, a post-loading pluton, the Golden Horn, displays original emplacement structure indicating growth by stacking of horizontal pluton sheets (Fig. 12-18) —offering a potential emplacement model for the older plutons of the Crystalline Core.

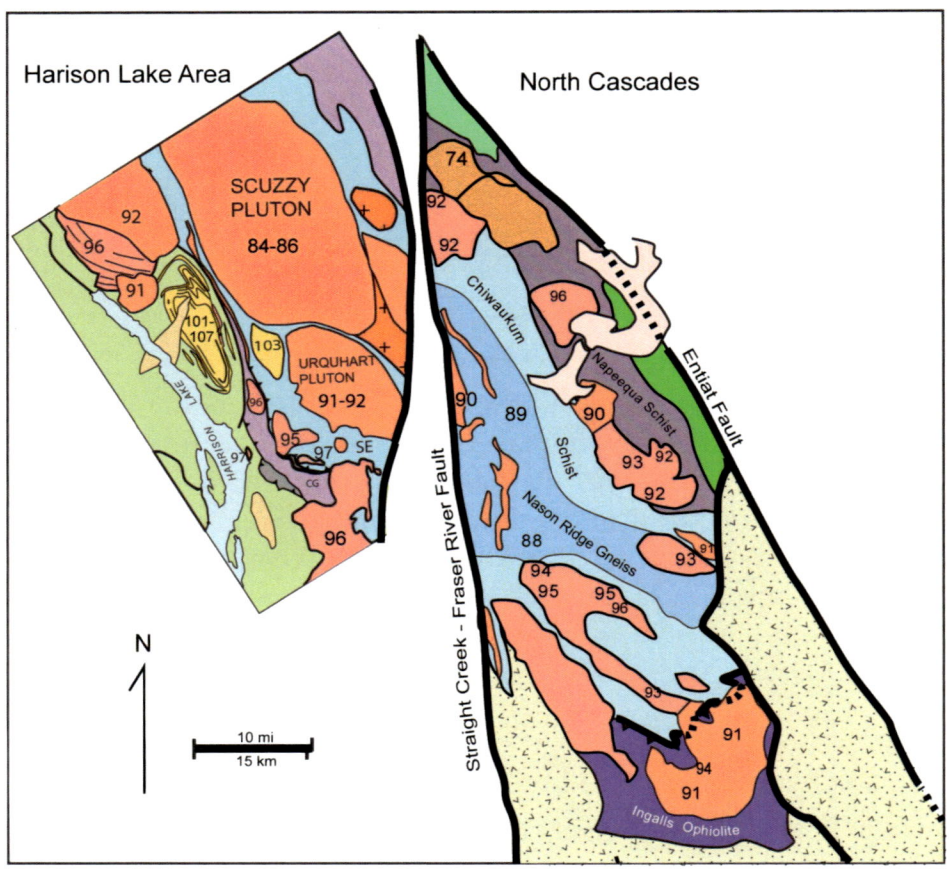

Fig. 12-35. Relative positions of plutons of the Harrison Lake area and North Cascades prior to displacement on the Straight Creek - Fraser River Fault. The Nason Ridge zone in the Cascades could be the floor of plutons similar to the Scuzzy and Urquhart.

Summary

The Cascades Crystalline Core is complicated. Many geologists have worked here and all-told have created a great amount of field and laboratory data. From these studies, and Canadian geology, we know that originally shallow country rock terranes were accreted by obduction at the margin of North America at about 170 Ma. The terranes were invaded by plutons in the Cascades region and contiguous Harrison Lake area of British Columbia from about 107 to 50 Ma. The terranes and plutons were deeply buried and then uplifted at different times in various localized zones within the overall Coast Plutonic Complex.

The cause of high-pressure metamorphism in the Cascades is debated. It basically comes down to structural contraction vs. magma loading, producing the great crustal thickening. Relevant thrusts are not found. Extension of Harrison Lake geology into the Cascades favors magma loading. But the metamorphic fabric of plutons and country rock in the Cascades can be interpreted as indicating structural thickening by bulk shortening—there is clearly scope for more research.

PART IV — MOUNTAINS WASHED AWAY, THEN RESURRECTED

CHAPTER 13

CHUCKANUT AND SWAUK FORMATIONS

The Chuckanut and Swauk formations mark the demise of the mountains built by the immense crustal displacements described for igneous and metamorphic rocks of the previous chapters. These sedimentary formations are mainly river-deposited sands (Fig. 13-1), defining a broad alluvial plain that extended from eastern Washington across the older eroded metamorphic and plutonic rocks of the Cascades and out to the San Juan Islands (Figs. 10-4 & 12-3). Remarkably, these deposits that were laid down in a tropical climate at very near sea level are now found in the high Cascades (Figs. 13-2). The time-span of Chuckanut - Swauk deposition is dated by detrital zircons at ~60 to 50 Ma. By 60 Ma, apparently much of the Cascades was beveled off by erosion and covered by the Chuckanut-Swauk alluvial plain. But the mountains rose up again—next chapter.

There is much to be said about the Chuckanut - Swauk formations, including the vastly different climate indicated by tropical flora and fauna. But our mission is mountain building, and in the next chapter we move on to thinking about how the mountains that were gone by 60 Ma became resurrected.

Fig. 13-1. Arkose sandstone of the Chuckanut Formation, exhibiting river-deposited cross-beds, along the coast near Larrabee Park.

Fig. 13-3. Palm frond in the Chuckanut Formation along Mt. Baker highway. John Feltman photo.

Fig. 13-2. Trunk of a palm tree in growth position in the Chuckanut Formation high in the Cascades (~6500' el.), near Bacon Peak (Fig. 10-4). David Bleam is dressed for the conditions. Ralph Haugerud photo.

CHAPTER 14

RAIN SHADOW

In the previous chapter, we saw that the mountains were beveled by erosion about 60 million years ago—the age of the near sea-level deposits of the overlying Chuckanut and Swauk formations. But, we have mountains now. As a consequence, Eastern Washington is arid—it is in a rain shadow. When and how did the mountains come back?

When the modern day mountains rose up is the easier part of this question. Two lines of evidence are at hand. One is fossil distribution. From the time of the Chuckanut - Swauk formations until about 8 Ma, fossil plants in eastern Washington are testimony to a wet climate and lush landscape (Fig. 14-1). There was little in the way of mountains to interrupt the east-moving rain clouds. Apparently, the Cascade Mountains were subdued from 60 to 8 million years ago. The same is true for the B.C. Coast Mountains.

Additional and confirming evidence for the prolonged period of minimal Cascades/Coast mountains

Fig. 14-1. Petrified tree trunks of the Ginko forest along the Columbia River at Vantage, Washington. The Ginko trees, and a great variety of other plants, grew in a lush setting in eastern Washington at a time about 15 Ma, before uplift of the Cascades when the rain shadow took hold. This is a time also of great eruption of flood basalts, seen in the background. Thanks to the basalts, the Ginko forest was buried, petrified, and preserved. Floodwaters, from melting continental glaciers, poured across the Columbia Plateau at the end of the Pleistocene epoch ~ 15,000 years ago, eroding the volcanic rock and exposing the petrified wood. Peggy Thompson photo.

comes from fission track dating. Uranium in zircon gives an age measure of the zircon by the proportion of uranium to lead, as we have seen in Chapter 3. The zircon can also give an age of when it came to the earth's surface, i.e. when mountain-building rocks were at the surface. Uranium in zircon is always giving off radiation. The emitted energy makes tracks in the zircon crystal. When the zircon crystal is warm (T> 175C), recrystallization removes the tracks. But when the temperature is less, the tracks are preserved, and the track density becomes a function of time. With a microscope, the diligent scientist can count tracks and get an age of how long since the rock was buried. For the Cascades/Coast mountains the zircon tracks indicate it hasn't been that long. The result is the same as that given by vegetation for the beginning of the rain-shadow effect: the mountains came up "only" about 8 million years ago.

We can understand that the original destruction of the mountains, by ~ 60 Ma, came about by erosion. But a *big question* arises in understanding *why* the mountains came back up at 8 Ma.

Interestingly, the Sierra Nevada mountains exhibit a similar history. The bedrock formed as a major magmatic arc, pluton ages are mostly in the range of 125—85 Ma, the mountains were eroded flat shortly after plutonism ceased, and the mountains were uplifted at about 8 Ma. Whatever caused the Sierras to come up is probably the same for the Cascades/Coast mountains batholith.

A hypothesis for this process was proposed by Mihai Ducea and Jason Saleeby (1998) with reference to the Sierra Nevada Batholith. The idea is that a dense cumulate root that formed at the base of the arc hung on long after the arc magmatism ceased, and neutralized the buoyancy effect of the pluton mass with respect to the floating equilibrium established in the asthenosphere (Chapters 2, 11). At about 8 million years ago, the heavy cumulate detached and sank; the upper part of the arc floated up by compensation in the asthenosphere. But we run into some difficulties with this explanation.

The Coast Plutonic Complex is an orogen that comprises a number of belts of country rock and plutons that had their own separate significant up and down motions, 10 to 20 miles, that reflect sagging during emplacement and rising during erosion. These motions happened at different times in the period from ~ 100 to 60 Ma, long before the ~8 Ma regional uplift. It would seem that the cumulate root would need to have been detached for these earlier uplifts to have occurred. In view of these separate large vertical displacements, it is difficult to envisage a regional, orogen-wide floor that maintained continuity and held on to it's cumulate basement until the much more recent uplift. If the uplift is connected to the growth of individual plutons, the simultaneous happening of uplift along both the Sierra and CPC seems illogical. Maybe a more likely cause is some type of regional plate tectonic shift.

The northeast buttress on Slesse Peak—a notable Fred Beckey climbing route. The mountain is carved out of a complex of Chilliwack Batholith (~35 Ma) and Darrington Phyllite of the northwest Cascades thrust system. Location on Fig. 5-1. Peter Jewett photo.

PART IV — SUMMARY

CHAPTER 15

TECTONIC EVOLUTION

How can we summarize the complex continental growth process recorded in the North Cascades and Harrison Lake area of the B.C. Coast Mountains? I thought a chart might be useful. Fig. 15-1 plots the high points. Here is a brief narrative to accompany the chart:

In the early days, 400 to 100 million years ago, we track two assemblages of rocks that had separate histories. In assemblage **1** are terranes of the northwest Cascades thrust system that accreted somewhere south and brought with them a record of ocean travels and subduction zone metamorphism. Assemblage **2** includes terranes that are country rock in the Coast Plutonic Complex. These terranes accreted locally by obduction prior to plutonism.

Assemblage **1** includes the Chilliwack terrane that wandered about as an island arc in the ancestral ocean for 250 million years before attaching to the edge of North America at least by 170 million years ago. This arc terrane carried with it a piece of ancient Greenland, giving us some idea of the travel path. The Bell Pass Mélange hosts many unrelated rocks in a tectonic mixture. Notable are a large chunk of the earth's mantle, the Twin Sisters Dunite, and a chert-basalt assemblage of oceanic rock bearing fossils indicating an origin south of Asia. The Easton Suite has a record of crust generation at an ocean ridge system, followed eventually by deep subduction and interaction with the earth's mantle. All these terranes have counterparts along the continental margin in Oregon and California.

In assemblage **2** we have for one, the Quesnellia terrane, an oceanic island arc complex of igneous and sedimentary rocks, apparently a distant relative of the Chilliwack terrane based on fossils. But, the Quesnellia terrane was not subducted; it accreted by riding up over the continental margin (obduction) in the region where we see it now. Another in this group is the Bridge River terrane, long-lived ocean crust, also obducted. Locally associated with the Bridge River terrane is a third component of assemblage **2**, the Cadwallader island arc assemblage. Assemblage **2** terranes underlay the western North American landscape in our neighborhood, just before plutonism began.

The Chiwaukum Schist and Methow formations are derived from sedimentary and volcanic rocks, 160 to 100 million years old, deposited in a forearc basin on the seaward side of Quesnellia and the continental margin.

Major plutonism of the Coast Plutonic Complex began at about 170 Ma, establishing the onset of vast continental arc magmatism extending some 1000 miles from Washington to Alaska. Plutons, in shifting belts, intruded assemblage **2** terranes ongoing for about 120 million years. The amount of granitic magma that came into the upper crust of the earth is staggering: ~40,000 cubic miles over the length and time span of the arc.

Halfway into its lifespan, the Coast Plutonic Complex was sliced by a sinistral strike-slip fault which carried the northern part of the arc in British Columbia and Alaska south to be outboard of the Methow basin. The Methow sedimentary environment was then transformed from being a "forearc" to being a "backarc" basin. South oblique motion of the Farallon Plate at this time possibly caused this great displacement of the older Coast Plutonic Complex. Convergent motion of the Farallon Plate against western North America switched from south oblique to north oblique at about 100 Ma.

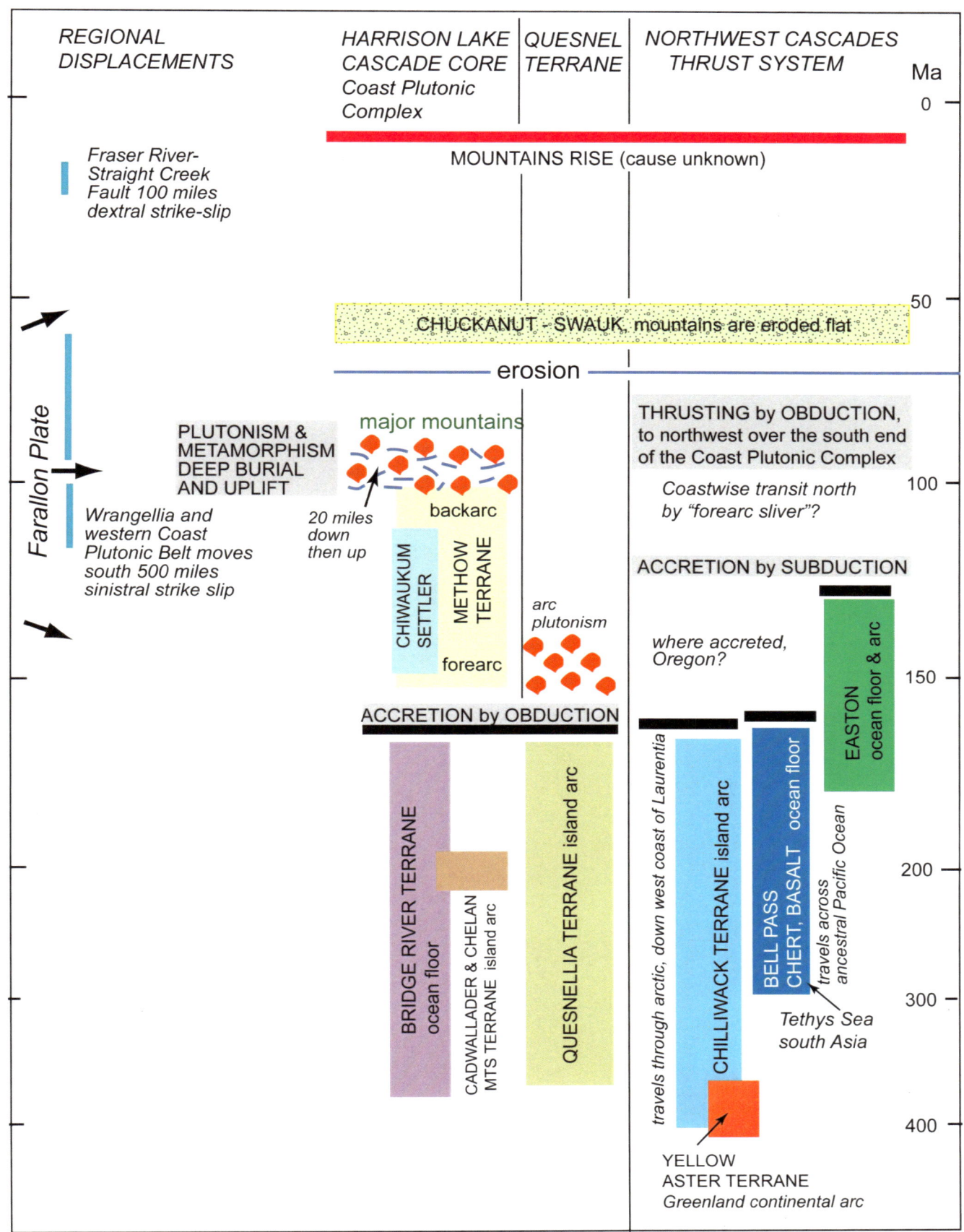

Fig. 15-1. Tectonic evolution of mountains of the North Cascades and Harrison Lake area. Farallon Plate arrows refer to the angle of convergence against North America, as in Fig. 1-5.

Major arc plutonism started up in the Harrison Lake - North Cascades area about 100 million years ago. Country rocks of the accreted terranes in proximity to the plutons were metamorphosed through a cycle of: 1) initial low-pressure and high-temperature, to 2) high pressure and high temperature, and then 3) uplift. Loading and unloading of the countryrock was on a scale of 10-20 miles of vertical displacement. Plutonism, metamorphism and loading occurred at different times in different belts, from ~95 to 65 Ma. The cause of loading is debated, but pluton emplacement in the upper crust is favored by me.

The nappes of the northwest Cascades thrust system, carried by the Farallon plate, accreted and were subducted along the continental margin south of the Pacific northwest, perhaps in Oregon and/or California. Then the terranes were exhumed and moved north. They reached their final resting point shoved up over the south end of the Coast Plutonic Complex. All this happened during the time of plutonism and high-grade metamorphism in the Coast Plutonic Complex.

Deeply buried rocks were uplifted and eroded by 60 Ma, when river sands of the Chuckanut and Swauk formations covered much of the area at near sea-level. The region remained at this level until about 8 Ma when the present day mountains were uplifted. The cause of this most recent uplift remains a mystery.

My hope for this book is that both the mountain traveler and the armchair reader will find stimulation in the dramatic mountain building events documented for the Cascades and Coast Mountains: plate collisions, great burial and uplift, and massive invasion of magma. The mountain geology is on display, and the book can be a guide for the layman to understand tectonic history underfoot and in view.

Mapping crew, Paul Pittman, Matt Nelson, and Kirsten Swanson, geared up for a traverse of the Mt. Mason Pluton, 1997.

GLOSSARY

accretion. Process of adding of rock masses to the continental margin by plate tectonics.

accretionary wedge. Thick wedge-shaped zone of accreted material at the continental margin, where rock materials on a converging oceanic plate are scraped off and added to the continent.

actinolite. A type of amphibole, green; common in low-grade metamorphic rocks.

Al. Abbreviation for the chemical element aluminum.

Al-in-hornblende. The aluminum content in hornblende amphibole gives a measure of pressure of formation.

aluminum silicates. The three polymorphs of Al_2SiO_5: andalusite, kyanite, sillimanite.

albite. Sodium-rich plagioclase feldspar: $NaAlSi_3O_8$.

alluvial. Refers to stream-deposited sediment.

amphibole. Dark, elongate mineral common in coarse-grained igneous rocks; rich in calcium, iron, magnesium, silicon.

amphibolite. Medium- to high-grade metamorphic rock, typically derived from basalt; rich in hornblende.

andalusite. One of the aluminum silicate minerals—see above.

andesite. Type of volcanic rock, fine-grained and dark grey, typically with light-colored feldspar crystals; common rock in arc volcanoes.

anticline. An arch type structure in folded rock.

aragonite. Form of calcium carbonate, $CaCO_3$. The usual form is calcite, which is the common mineral of limestone. With great depth of burial (>12 miles), calcite changes to the polymorph aragonite due to the increased confining pressure.

arc. Belt of igneous intrusions and volcanoes formed above a subduction zone, where the earth's crust dives down and releases fluids at a depth of 50 to 70 miles that rise and cause melting in the over-riding plate.

arkose. Sandstone made largely of quartz and feldspar grains. This sand typically comes from erosion of granite.

argillite. Sedimentary rock made of mud and silt; harder and more massive than shale.

asthenosphere. Deep layer in the earth, generally 20 to 50 miles below the earth's surface, distinguished by slow velocity of earthquake waves. This zone is close to melting, is plastic. More rigid earth layers above slide across the asthenosphere, allowing plate tectonic transport. When heavy loads are placed on the earth's crust, such as glaciers or sheets of faulted rock, the asthenosphere sinks under the load, providing a type of floating equilibrium.

atomic structure. Distinctive latticework of structure in a mineral built up of atoms of the constituent elements. In the case of $CaCO_3$, the atoms of calcium, carbon and oxygen can be organized to form either the mineral calcite, or alternatively aragonite. The two minerals have the same chemistry, but different atomic structure. The aragonite crystal lattice is more compact, reflecting the high-pressure conditions under which it formed.

aureole. With reference to the country rock intruded by plutons, the term aureole applies to the region of contact metamorphic rocks affected by the pluton.

backarc basin. A sedimentary basin flanking the arc on the side opposite of the subduction zone.

Baja B.C. A concept, based on paleomagnetism, that older rocks (>70 Ma) of western Washington and British Columbia came from the region of Baja California, moved by strike-slip faulting along the continental margin.

Barrovian. A term referring to a sequence of metamorphic zones recognized in Scotland by George Barrow.

basalt. Dark, fine-grained volcanic rock, very common. Basalt magma mostly comes from partial melting of mantle rocks (peridotite) 10 to 50 miles deep in the earth.

Benioff zone. Spatial pattern of earthquakes where an oceanic plate runs into, and plunges under, another plate, either oceanic or continental. The starting points (foci) of these earthquakes lie roughly in a dipping plane, and mark the array of faults caused by subduction of the oceanic plate.

belemnite. A bullet-shaped fossil bone from an animal related to modern day squid; common in the Nooksack Formation; Mesozoic in age (250-65 Ma).

biotite. Brown to black mica, iron and magnesium-rich. Common in granite.

blueschist. A blue foliated metamorphic rock formed at conditions of high pressure - low temperature, indicating a subduction zone. Sodic amphibole give the rock its blue color.

breccia. Fragmental rock made of relatively coarse (> 1/10 in.), angular chunks. Igneous breccia forms by explosive volcanic action. Sedimentary breccia typically forms from talus at the base of a cliff. Fault breccia forms by breakage in a fault zone.

buoyant uplift. Crustal rocks uplifted by a floating equilibrium in the plastic asthenosphere when a load on the earth's surface is diminished, as when mountains are eroded; the rock floats somewhat like a block of wood in water.

CPC. Coast Plutonic Complex; a 1000 mile long belt of plutons and country rock extending from Washington to Alaska.

calcite. Carbonate mineral ($CaCO_3$). For our purposes, this is the mineral that makes up limestone and is derived from fossil shells; change to the polymorph aragonite occurs with deep burial and metamorphism.

Cascades Crystalline Core. Plutonic and metamorphic rocks in the North Cascades; the south end of the CPC.

cathode luminescence. Luminescence caused by irradiation with electrons; brightens minerals viewed with the electron microscope.

chert. Rock made of fine-grained quartz or opal. In the Pacific Northwest, chert comes from accumulations of siliceous plankton—radiolaria, in a deep ocean environment.

chlorite. Green platy metamorphic mineral, altered from igneous pyroxene or amphibole. Iron-magnesium-aluminum silicate.

chromite. A black metallic mineral, $(Fe, Mg) Cr_2O_4$, common in dunite and peridotite.

clast. Fragment in a sedimentary rock, such as a sand grain or cobble.

clastic. Textural term applied to sedimentary rock that is made of fragments; for example, shale, sandstone, conglomerate, breccia.

cleavage. Planar fabric that penetrates a metamorphic rock—like pages in a book. Is caused by squeezing of the rock to align platy minerals, such as mica. Also, a term for a plane of weakness within a crystal, related the crystal lattice structure.

clinopyroxene. Diopside or augite, monoclinic in structure, rich in iron and magnesium.

Cocos Plate. Oceanic plate that converges against, and is subducted under, Mexico and Central America.

confining pressure. Pressure coming equally from all sides, related to depth of burial.

conglomerate. Clastic sedimentary rock made up of pebbles and cobbles that are at least somewhat rounded, distinguishing conglomerate from breccia.

contact aureole. Metamorphic zone along the margins of a pluton caused by heat emanating from the igneous body.

continental arc. Belt of igneous rocks developed in a continental land mass, fed from an underlying subduction zone.

continental drift. Hypothesis that continents move about; the concept dating to the early 1900s, and earlier. In modern times we recognize that continents have moved by plate tectonics.

cordierite. A metamorphic mineral formed at relatively high temperature and low pressure—$Mg_2Al_4Si_5O_{18}$.

country rock This term refers to the rocks that host plutons. The plutons intruded into the country rock, causing metamorphism of the country rock, and in some places displacement making room for the plutons.

craton. An older part of a continent that no longer is involved in mountain building.

crinoids. Marine animals, commonly in a structure that resembles plants. Important fossil in the Chilliwack Fm.

crossite. Blue, sodium-rich amphibole that formed at high pressure in a subduction zone.

crystal lattice. Three dimensional configuration of atoms (or ions) bonded together, constituting the internal structure of a crystal.

cumulate. Coarse-grained dark igneous rock that formed by crystallization, settling, and accumulation of mineral grains on the bottom of the magma chamber, commonly a gabbro or peridotite.

dacite. Light-colored volcanic rock. Similar to rhyolite but lacks visible quartz in hand sample.

decay rate. Time it takes for a radioactive substance to transform into a daughter product; for example, uranium giving off radiation and changing to lead. Usually expressed as the half-life—how long for half the radioactive substance to be destroyed..

detrital zircon. Zircon occurring as sedimentary grains.

derivative magma. Magma, molten rock, that is changed in composition from the original magma by subtraction of crystallized parts.

diapir. An intruding pluton that ascends through the earth's crust as a streamlined blob, comparable to a bubble rising in water. The intruded country rock flows around the pluton.

dikes. Sheet-like intrusions of igneous rock intruded across structure in the country rock.

diorite. Coarse-grained igneous rock, not as dark as gabbro, usually has visible grains of light-colored feldspar, and dark amphibole.

ductile deformation. Process of rock deformation by bending or flowing, not breaking.

dunite. A rock made nearly all of olivine.

Earth's core. Innermost part of the earth; made of an iron-nickel alloy.

Earth's mantle. Zone in the earth between the core and crust, about 1800 miles thick, made of magnesium-iron silicate minerals.

eclogite. Metamorphic rock characterized by garnet and omphacite (sodium-rich pyroxene), formed at high pressure and moderate to high temperature.

electron microscope. Microscope that shines electrons on the sample of interest to obtain an image resolving detail on a very fine scale.

entropy. A thermodynamic property of a substance related to the degree of order. Water has higher entropy than ice. In minerals, higher entropy forms are stable at higher temperature.

epidote. Metamorphic mineral distinguished by its pistachio green color. Calcium-aluminum silicate.

Farallon Plate. Ancient crustal plate of the ancestral Pacific Ocean. Most of the plate has been subducted along the western margin of North and South America. Remnants are the Cocos, Nazca, and Juan de Fuca plates.

fault. Fracture plane in rocks along which one side has slid relative to the other side.

Fe. Iron.

feldspar. Light-colored Na-K-Ca-silicate mineral common in igneous rocks, and in arkosic sandstone.

fission track. A microscopic trail in a crystal made by natural radiation from uranium in the crystal.

flood basalts. Broad sheets of basalt, many miles across, emanated as lava from deep earth cracks above a "hot spot".

fold. Bend in rock. Folds can be tight or open, minute or mountain size.

forearc. Geologic setting in front of the arc, on the side of the subducting plate.

forearc sliver. A tract of earth crust lying between the active arc and the subduction zone.

foreland basin. A sedimentary basin between the active arc and the subduction zone.

fractionation of magma. Process of a parent magma crystallizing into different rock types. Early-formed crystals typically have a different composition than the parent magma. If the early-formed crystals are removed from reacting with the remaining magma, then the original magma has fractionated into different rock types.

Franciscan complex. An assemblage of ocean-derived rocks in the California Coast Range that have been metamorphosed in a subduction zone to blueschist and eclogite.

gabbro. A coarse-grained igneous rock composed of the minerals: plagioclase feldspar and pyroxene, and sometimes olivine.

garnet A hard, usually red, equant mineral common in metamorphic rocks. Ca-Fe-Mg aluminum silicate.

geochronology. Science of determining rock ages.

geomagnetic time scale. The age of rocks determined from their pattern of paleomagnetic pole reversals.

geothermal gradient. Rate of increase of temperature with depth in the earth.

gneiss. A strongly metamorphosed rock in which light and dark colored mineral have separated into lenses and layers.

granite. Coarse-grained, light-colored igneous rock, specifically made mostly of quartz and potassium feldspar.

granitic. A general descriptive term for coarse-grained, quartz-feldspar igneous rocks.

graphite. Made of carbon; the low-pressure polymorph of diamond.

greenschist. Low-grade foliated metamorphic rock consisting largely of chlorite, epidote, and actinolite.

greenstone. Fine-grained, non-foliated, low-grade, green metamorphic rock.

greywacke. Sandstone composed of dark sand grains of mainly volcanic rock and chert. Mud matrix is common.

heavy liquid. Chemical with greater density than many minerals, useful for separating heavier from lighter minerals.

hornblende. Type of amphibole. Iron-magnesium silicate, dark and elongate, found commonly in granitic rocks.

hot spot. Localized zone of hot upwelling of mantle material that causes melting of peridotite to produce basalt. A good example is the Hawaiian hot spot.

igneous rock. Rock that forms by cooling and solidification of melt (magma).

injection complex. Multiple dike intrusions commonly observed in a single outcrop.

ion. An element with an electronic charge, due to either more or fewer electrons than protons. e.g Fe^{2+}.

island arc. Chain of oceanic volcanic islands developed over a subduction zone.

Isostatic equilibrium. The floating equilibrium of plates of the lithosphere on the denser plastic asthenosphere.

isotherm. A delineation of a uniform temperature.

isotope. Many chemical elements have versions with different make-ups of the atom in terms of number of neutrons. Thus, the carbon 12 isotope has 6 protons and 6 neutrons; whereas the carbon 14 isotope has the same number of protons, but 8 neutrons. Carbon 12 is stable; carbon 14 changes (decays) to nitrogen.

Juan de Fuca Plate. Small ocean plate off the coast of the Pacific Northwest. A remnant of the ancestral Farallon Plate, mostly subducted. The Juan de Fuca Plate is actively subducting and gives rise to the Cascade volcanoes.

kinematics. Movements in rocks; folding, faulting etc.

kyanite. An aluminum silicate polymorph; Al_2SiO_5.

laser spot. Refers to a tiny pit (~.03 mm) on a zircon grain created during age analysis.

Laurentia. Name given to ancestral North America, as in Precambrian times, before the addition of all rocks west of the Rocky Mountains.

limestone. Sedimentary rock made of calcite, generally derived from fossil shells.

lithosphere. An outer part of the earth consisting of the uppermost mantle and the earth's crust.

low velocity zone. The asthenosphere zone within the earth's mantle that is close to melting, where S-type earthquake waves are slow. The plastic nature of the asthenosphere accommodates the sliding of overlying plates of the lithosphere, and the floating equilibrium of roots of crustal loads, such as mountains.

Ma. Millions of years ago.

Mn. Manganese.

Mg. Magnesium.

magnetite. Black iron oxide mineral that will stick to a magnet; Fe_3O_4.

magma. Molten rock.

magmatic arc. Elongate belt of igneous rocks, plutons and volcanoes, developed over a subduction zone.

magma chamber. A zone inside the earth, in the crust or mantle, where magma has accumulated.

magmatic foliation. Planar igneous fabric in a pluton formed by the alignment of tabular minerals while the pluton was still partially molten.

mass spectrometer. Instrument that distinguishes particles of different mass. Can measure different isotopes and is thus critical for some types of isotopic age-dating, e.g. uranium-lead in zircon.

mélange. A term applied to a rock that is a chaotic structural mixture of various rock types.

metamorphic rock. Type of rock formed where a sedimentary or igneous rock is buried in the earth or intruded by a pluton so that increased heat and pressure change the minerals and rock textures.

migmatite. A rock, typically on outcrop scale, that is an intimate mixture of country rock and igneous rock, formed at high temperature.

miogeocline. A broad apron of sedimentary rocks along the western passive margin of Laurentia (~600-370Ma) before the Cordillera developed.

mineral stability. Relates to whether a mineral has a force upon it to change. Depending on the temperature, pressure and composition of the surroundings, an unstable mineral can melt, dissolve in fluid, or change to another mineral.

moho. Boundary between the earth's crust and underlying mantle.

monazite. A rare phosphate mineral that contains trace amounts of uranium, allowing for U-Pb dating.

mudstone. Dense, fine-grained sedimentary rock formed by compaction of mud.

muscovite. A common mica in metamorphic rocks; platy, shiny, translucent, colorless to light grey.

nappes. Large, miles wide, sheets of rock that have been thrust over other rocks or the landscape.

Nazca Plate. Remnant of the ancestral Farallon Plate that is currently being subducted under the west coast of South America.

North American Cordillera. Tract of land extending approximately from the Rocky Mountains to the west coast in breadth, and from Alaska to Mexico in length. This domain includes sedimentary rocks along the western margin of Laurentia and much other rock added to the continent by way of terrane accretion, plutonism and volcanism.

ocean crust. Consists of ocean-floor basalt generated at an ocean ridge system or an ocean island hot spot. Older ocean crust has a sediment overlay, typically of chert.

ocean island. Volcanic island generated over a hot spot within the oceanic plate. A long-lasting hot spot will create a chain of islands as the ocean crust moves along over the magma source (Hawaiian chain).

ocean ridge. Ridge system formed where ocean plates move apart. Basalt magma is generated in the upper mantle, intrudes upward to the ridge where it crystallizes, and adds to the edge of newly formed ocean crust.

ocean trench. Large-scale depression formed where the ocean plate dives down into a subduction zone. The trench is coupled with an arc formed over the subduction zone.

omphacite. A greenish pyroxene that is an important component of eclogite, formed deep in a subduction zone.

olivine. Magnesium silicate mineral that is a major component of the upper mantle; abundant in the rocks peridotite and dunite.

ophiolite. An on-land sequence of rocks that is an uplifted oceanic formation. Some ophiolites are crustal sections of an island arc, other ophiolites represent ocean crust as generated at the ridge system.

orogeny. Large scale process of mountain building, by way of magmatism, metamorphism, and tectonism.

orthoclase. Common feldspar in granite. Potassium aluminum silicate.

orthogneiss. An igneous rock that has been strongly metamorphosed, creating a fabric of layers or lenses of light and dark mineral grains.

outcrop. Exposure of rock.

Panthalassa. A global scale ocean surrounding the supercontinent Pangea.

Pangea. A super continent. All the earth's continents were together at ~250 Ma.

paragneiss. A gneissic rock formed by metamorphism of sedimentary deposits.

passive margin. Continental margin not affected by plate collision; no volcanic arcs, limited earthquakes.

PBR Pabst Blue Ribbon.

peridotite. Rock made of olivine and pyroxene, prevalent in the upper mantle.

phyllite. Fine-grained, low-grade platy metamorphic rock derived from mudstone or siltstone.

pillow basalt. Basalt formed with pillow structure that develops as the magma solidifies under water.

plagioclase. Common feldspar in igneous rocks. Sodium, calcium, aluminum silicate.

planar fabric. Planes of alignment of platy minerals, typically in a metamorphic rock. Foliation.

plastic. Condition of a rock where it can flow as a solid.

plate tectonics. Theory of dynamic mechanisms in the outer earth. The earth, from the crust into the upper mantle, is divided into plates, each of which is up to tens of miles thick and thousands of miles in breadth. The plates slide about on a slippery substrate of the asthenosphere, and their motions cause earthquakes, volcanoes, and virtually all other large-scale movements of the earth's crust.

pluton. A body of intrusive igneous rock that is relatively coarse-grained and crystallized slowly well below the earth's surface.

point-counting. A method of estimating the proportions of different minerals in a rock by tabulating mineral species at hundreds of points on a grid viewed through a microscope.

polarized light. Light that vibrates in a single plane. Used in microscopy.

polymorph. A solid of a certain chemical formula that can take different structural forms. Calcite and aragonite, both $CaCO_3$, are polymorphs. They have different internal structure.

Precambrian. Period of geologic time from the beginning of Earth to 540 Million years ago.

prehnite. Green metamorphic mineral. Calcium-aluminum silicate.

protolith. The original rock, prior to being metamorphosed.

provenance. Source area, as for example, the rock or region from which grains in a sandstone were derived.

pumpellyite. Green metamorphic mineral. Calcium-iron-magnesium-aluminum silicate.

pyrite. Iron sulfide mineral.

pyroclastic. Fragmental texture in an igneous rock produced by an explosive volcanic eruption.

pyroxene. Iron-magnesium silicate mineral common in gabbro and peridotite.

quartz. Silicon dioxide, common in rocks, resistant to weathering.

Quesnellia. A terrane in south-central British Columbia consisting mainly of island arc igneous rocks and associated volcanic sediments, formed at sea marginal to North America ~370-190 Ma.

radioactive decay. Process of a radioactive element, e.g. uranium, emitting radiation and breaking down to a "daughter" product—lead in the case of uranium.

radiometric age. Age of a mineral determined by measuring the amount of radioactive parent and daughter product of radioactive decay.

radiolaria. Tiny, siliceous, planktonic organisms. Mass accumulation of dead radiolaria raining down on the sea floor is the raw material for creating chert.

remnant arc. Island arc that is no longer an active volcano, having lost its connection to the subduction zone magma source as a consequence of the subducting plate stepping back, away from the arc.

ribbon chert. Chert beds interlayered with softer, recessive shale beds.

Rodinia. An ancient grouping of continents. At different times in earth history all the continents have been joined together into one giant landmass. Rodinia is one such grouping, existing about from about 1100 to 750 million years ago. It split into numerous continents, including Laurentia, the ancestral North America. The continents all got back together at about 300 million years ago, forming the supercontinent Pangea, which split about 200 million years ago to yield early forms of the continents we have now.

root zone. The area where a thrust sheet of rock has broken out of its position prior to thrusting.

schist. A metamorphic rock characterized by strong foliation and grains easily seen without magnification.

seamount. An extinct ocean island volcano that, due to plate motion, has moved off its hot spot magma source.

sea-floor spreading. Spreading of oceanic plates away from an ocean ridge system. The plate grows by addition of basalt at the ridge, then moves away toward a subduction zone sliding on a gradual downhill gradient along the top of the asthenosphere.

sedimentary rock. Rock formed at the earth's surface by accumulation and hardening of various substances: weathered material (sand etc.), organisms (plant or animal), or by chemical precipitates (salt or other).

Seismic waves. Earthquake energy propagated by waves of earth motion.

semi-schist. Like a schist, but with some relict sedimentary or igneous grains.

serpentinite. Rock made of the mineral serpentine, a magnesium-rich silicate that forms by alteration of olivine in peridotite.

shear pressure. Directed stress on a rock causing a displacement by sliding of one part past another. This is in contrast to confining pressure which squeezes a rock in all directions, making it smaller.

Si. The element silicon.

silica. Quartz.

siliceous. Rock rich in silica.

sillimanite. A polymorph of Al_2SiO_5. Relatives are andalusite and kyanite.

sills. Sheets of igneous rock intruded parallel to structure in the country rock (e.g. sedimentary bedding).

staurolite. A brown, medium grade metamorphic mineral, commonly occurring as large crystals in a mica schist. $Fe_2Al_4Si_4O_{22}(OH)$.

Stikinia. An accreted terrane in north-central British Columbia; similar and related to the Quesnellia terrane.

stoping. Process of a pluton making room for itself, while rising the earth's crust, by breaking off roof fragments of the overlying country rock.

strike and dip. Measurements that give the orientation of a planar geologic surface, such as sedimentary bedding. The strike is the compass direction of a horizontal line lying in the plane. The dip line is at right angles to the strike, it follows the fall-line down the surface, and the dip reading is the angle between the dip line and horizontal.

strike-slip. Horizontal displacement on a fault parallel to the strike of the fault.

subduction zone. Region where ocean crust collides with another plate and plunges down into the mantle.

talc. A very soft magnesium silicate mineral formed by metamorphism of olivine-rich rocks: peridotite and dunite.

tectonics. Relates to large-scale crustal movements: plate interactions, mountain building, etc.

terrane. Large crustal block of rock that has moved from a place of origin to be faulted into the place where we find it now. This rock is "exotic" to the land in which it is found.

Tethys Sea. Part of the ancestral Pacific Ocean between Asia and eastern Africa, about 250-150 million years ago.

thermobarometry. Measurements of the temperature and pressure at which rocks formed.

thermodynamics. A branch of physics and chemistry that allows calculation of pressure-temperature mineral stability from various parameters, such as volume, energy, entropy.

thrust sheet. Nearly horizontal slab of rock that has been carried over or under another block of land.

tonalite. Fine-grained igneous rock composed chiefly of plagioclase feldspar and quartz.

tuff. Volcanic ash.

ultramafic. Describes a rock rich in magnesium and iron bearing minerals, such as olivine and pyroxene. The common ultramafic rocks are peridotite and dunite.

under-plated. Rock materials added to the continental margin by accretion (sticking on) of thrust sheets to the bottom of the over-riding plate in a subduction zone.

volcanic rock. Erupted igneous rock.

weathering. Process of natural breakdown of rocks and minerals. Chemical weathering occurs by dissolution in surface waters. Mechanical weathering is rock break-up caused by a variety of processes: frost wedging, transport over cliffs, bouncing in river beds, pounding in surf zones, etc.

Wrangellia. Large terrane accreted onto the outermost northwest margin of North America, extending from Vancouver Island to Alaska. The terrane has a long history of formation as ocean crust, 350-150 Ma, including some parts developed as oceanic plateaus and other parts as island arcs.

zircon. $ZrSiO_4$. A common, but very minor, mineral in igneous rocks and in sediments eroded from igneous rocks. Trace amounts of uranium in zircon allow radiometric age determination.

REFERENCES

1. Arthur, A. A., Smith, P. L., Monger, J. W. H., and Tipper, H. W. 1993. Mesozoic stratigraphy and Jurassic paleaontology west of Harrison Lake, southwestern British Columbia. Geological Survey of Canada Bull. 441, 62p.

2. Bradley, R., Karlstrom, K. E., Hawkins, D. P., and Williams, M. L. 1996. Tectonic evolution of Paleoproterozoic rocks in the Grand Canyon: Insights into middle-crustal processes. Geol. Society of America Bulletin, 108: 1149-1166.

3. Brown, E. H., Wilson, D. L., Armstrong, R. L., and Harakal, J. E. 1982. Petrologic, structural, and age relations of serpentintite, amphibolite, and blueschist in the Shuksan Suite of the Iron Mountain-Gee Point area, North Cascades, Washington. Geological Society of America Bulletin, 93: 1087-1098.

4. Brown, E. H., Blackwell, D. L., Christenson, B. W., Frasse, F. I., Haugerud, R. A., Jones, J. T., Leiggi, P. A., Morrison, M. L., Rady, P. M., Reller, G. H., Sevigny, J. H., Silverberg, D. S., and Ziegler, C. B. 1987. Geologic map of the northwest Cascades, Washington. Geol. Society of America Map and Chart Series MC-61, 1 sheet, 1:100,000.

5. Brown, E.H. 1987. Structural geology and accretionary history of the Northwest Cascades system, Washington and British Columbia. Geological Society of America Bulletin v. 99: 201-214.

6. Brown, E. H., and Talbot, J. L. 1989. Orogen-parallel extension in the North Cascades crystalline core. Tectonics 8: 1105-1114.

7. Brown, E. H., and Walker, N. W. 1993. A magma loading model for Barrovian metamorphism in the southeast Coast Plutonic Complex, British Columbia and Washington. Geol. Society of America Bulletin, 105: 479-500.

8. Brown, E. H., Cary, J. A., Dougan, J. D., Dragovich, J. D., Fluke, S. M., and McShane, D. P. 1994. Tectonic evolution of the Cascade crystalline core in the Cascade River area, Washington. WA Div. of Geol. Ear. Res., Bull 80: 93-113.

9. Brown, E. H., Talbot, J. L., McClelland, W. C., Feltman, J. A., Lapen, T. J., Bennett, J. D., Hettinga, M. A., Troost, M. L., Alvarez, K. M., and Calvert, A. T. 2000. Interplay of plutonism and regional deformation in an obliquely convergent arc, southern Coast Belt, British Columbia. Tectonics, 19: 493-511.

10. Brown, E. H., and McClelland, W. C. 2000. Pluton emplacement by sheeting and vertical ballooning in part of the southeast Coast Plutonic Complex, British Columbia. Geol. Society of America Bulletin, 112: 708-719.

11. Brown, E. H., and Gehrels, G. E. 2007. Detrital zircon constraints on terrane ages and affinities and timing of orogenic events in the San Juan Islands and North Cascades, Washington. Can. Journal of Earth Sciences, 44: 1375-1396.

12. Brown, E. H., Gehrels, G. E., and Valencia, V. A. 2010. Chilliwack composite terrane in northwest Washington: Neoproterozoic-Silurian passive margin basement, Ordovician-Silurian arc inception. Canadian Journal of Earth Science, 47: 1347- 1366.

13. Burchfiel, B. C., Cowan, D. S., and Davis, G. A. 1992. Tectonic overview of the Cordilleran orogen in the western United States. *In* The Cordilleran Orogen: Conterminous U.S., *Edited by* B. C. Burchfiel, P. W. Lipman and M. L. Zoback. Geological Society of America, Boulder, Colorado, G-3: 407-479.

14. Colpron, M., Nelson, J. A., and Murphy, D. C. 2007. Northern Cordilleran terranes and their interactions through time. GSA Today, 17: 4-10.

15. Cordey, F., and Schiarizza, P. 1993. Long-lived Panthalassic remnant: the Bridge River accretionary complex, Canadian Cordillera. Geology, 21: 263-266.

16. Cordova J. L., Mulcahy, S. R., Schermer, E., and Webb, L. E. 2017. Initiation and early evolution of a subduction zone: T-T-D history of the Easton Metamorphic Suite, Northwest Washington State. Geol. Soc. Amer. Abst. v. 49 no.6.

17. Crawford, W. A., and Hoersch, A. L. 1972. Calcite - aragonite equilibrium from 50 degrees C to 150 degrees C. American Mineralogist, 57: 995-998.

18. Danner, W. R. 1970. Paleontologic and stratigraphic evidence for and against sea floor spreading and opening and closing oceans in the Pacific Northwest. Geol. Society of America Abstracts with Programs, v. 2, no. 2: 84-85.

19. DeBari, S. M., and Greene, A. R. 2011 Vertical stratification of composition, density, and inferred magmatic processes in exposed arc crustal sections. In: Arc-Continent Collision. D. Brown and P. D. Ryan editors. Springer: 121-144.

20. Dickinson, W. R., and Hatherton, T. 1967. Andesitic volcanism and seismicity around the Pacific. Science, 157: 801-803.

21. Ducea, M., and Saleeby, J. B. 1998. A case for delamination of the deep batholithic crust beneath the Sierra Nevada, California. International Geology Review, 40: 78-93.

22. Dungan, M. A., Vance, J. A., and Blanchard, D. P. 1983. Geochemistry of the Shuksan greenschists and blueschists, North Cascades, Washington: Variably fractionated and altered basalts of oceanic affinity. Contributions to Mineralogy and Petrology, 82: 131-146.

23. Eddy, M. P., Bowring, S. A., Umhoefer, P. J., Miller, R. B., McLean, N. M., and Donaghy. E. E., 2016. High-resolution temporal and stratigraphic record of Siletzia's accretion and triple junction migration from nonmarine sedimentary basins in central and western Washington. Geol. Soc. Amer. Bull. v. 128: 425-411.

24. Eddy, M.P., Bowring, S. A., Miller, R. B., and Tepper, J. H. 2016. Rapid assembly and crystallization of a fossil large-volume silicic magma chamber. Geology v. 44: 331-334.

25. Engebretson, D. C., Cox, A., and Gordon, R. G. 1985. Relative motions between oceanic and continental plates in the Pacific basin. Geological Society of America Special Paper 206.

26. Evans, B. W., and Berti, J. W., 1986. Revised metamorphic history for the Chiwaukum Schist, North Cascades, Washington. Geology v. 14: 695-698.

27. Ferry, J. M., and Spear, F. S. 1978. Experimental calibration of the partitioning of Fe and Mg between biotite and garnet. Contributions to Mineralogy and Petrology, 66: 113-117.

28. Friedman, R. M., and Armstrong, R. L. 1995. Jurassic and Cretaceous geochronology of the southern Coast Belt, British Columbia, 49° to 51° N. Geological Society of America Special Paper 299: 95-139.

29. Gardner, D. W. 2006. Sedimentology, stratigraphy, and provenance of the upper Purcell Supergroup, southeastern British Columbia, Canada: implications for syn-depositional tectonism, basin models, and paleogeographic reconstructions. Master of Science Thesis, University of Victoria, 73 pages.

30. Gatewood, M. P., and Stowell, H. H. 2012. Linking zircon U-Pb and garnet Sm-Nd ages to date loading and metamorphism in the lower crust of a Cretaceous magmatic arc, Swakane gneiss, WA, USA. Lithos, doi 10.1016: 128-142.

31. Gehrels, G. E., Woodsworth, G., Crawford, M., Andronicos, C., Hollister, L., Patchett, J., Ducea, M., Butler, R. F., Klepeis, K. A., Davidson, C., Friedman, R. M., Haggart, J. W., Mahoney, J. B., Crawford, W., Pearson, D., and Girardi, J. 2009. U-Th-Pb geochronology of the Coast Mountains batholith in north-coastal British Columbia: constraints on age and tectonic evolution. Geological Society of America Bulletin, 121: 1341-1361.

32. Gibson, H. D., and Monger, J. W. H. 2014. The Cretaceous-Cenozoic Coast-Cascade orogen: the Chilliwack Valley-Harrison Lake connection. The Geological Society of America Field Guide, Field Trip 411: 45. http://www.sfu.ca/~hdgibson/Research/Gibson&Monger_GSA-FieldGuide_411_2014.pdf

33. Haugerud, R. A., Morrison, M. L., and Brown, E. H. 1981. Structural and metamorphic history of the Shuksan Metamorphic Suite in the Mount Watson and Gee Point areas, N. Cascades, WA. Geol. Soc. Amer. Bull. 92: 374-383.

34. Haugerud, R. A., and Tabor, R. W. 2009. Geologic Map of the North Cascade Range, Washington. U.S. Geological Survey Scientific Investigations Map 2940

35. Haugerud, R. H., Van der Heyden, P., Tabor, R. W., Stacy, J. S., and Zartman, R. E. 1991. Late Cretaceous and early Tertiary plutonism and deformation in the Skagit Gneiss Complex, North Cascade Range, Washington and British Columbia. Geological Society of America Bulletin, 103: 1297-1307.

36. Haugerud, R. A., Brown, E. H., Tabor, R. W., Kriens, B. J., McGroder, M. F., 1994. Late Cretaceous and early Tertiary orogeny in the North Cascades. in Swanson, D. A.. And Haugerud, R. A., editors, Geologic field trips in the Pacific Northwest: v. 2, p. 2E 1 - 2E 53. Geol. Soc. America field guide.

37. Hill, D. P. 1969. Crustal structure of the island of Hawaii from seismic-refraction measurements. Bulletin of the Seismological Society of America, 59: 101-130.

38. Holdaway, M. J. 1971. Stability of andalusite and the aluminum silicate phase diagram. Amer. Jour. Sci., 271: 97-131.

39. Matzel, J. P., Bowring, S. A., and Miller, R. B. 2004. Protolith age of the Swakane Gneiss, North Cascades, Washington: evidence of rapid underthrusting of sediments beneath an arc. Tectonics, 23: 1-18.

40. Matzel, J. E. P., Bowring, S. A., and Miller, R. B. 2006. Time scales of pluton construction at differing crustal levels: Examples from the Mount Stuart and Tenpeak plutons, North Cascades. Washington. Geol. Soc. of America Bulletin, 118: 1412-1430.

41. McGroder, M. F. 1991. Reconciliation of two-sided thrusting, burial metamorphism, and diachronous uplift in the Cascades of Washington and British Columbia. Geological Society of America Bulletin, 103: 189-209.

42. Michel, J., Baumgartner, L., Putlitz, B., Schaltegger, U., and Ovtcharova, M. 2008. Incremental growth of the Patagonian Torres del Paine laccolith over 90 k.y. Geology. v.36: 459-462.

43. Miller, M. M. 1987. Dispersed remnants of a northeast Pacific fringing arc: upper Paleozoic terranes of Permian McCloud faunal affinity, western U.S. Tectonics, 6: 807-830.

44. Miller, R. B. 1985. The ophiolitic Ingalls Complex, north-central Cascade Mountains, Washington.. GSA Bull., 96: 27-42.

45. Miller, R. B., and Bowring, S. A. 1990. Structure and chronology of the Oval Peak batholith and adjacent rocks: Implications for the Ross Lake fault zone, North Cascades Washington. Geol. Soc. Amer. Bull., 102: 1361-1377.

46 . Miller, R. B., and Paterson, S. R. 1992. Tectonic implications of syn- and post-emplacement deformation of the Mount Stuart batholith for mid-Cretaceous orogenesis in the North Cascades. Can. Jour. of Earth Science, 29: 479-485.

47. Miller, R. B., and Paterson, S. R. 2001. Construction of mid-crustal sheeted plutons: Examples from the North Cascades, Washington. Geological Society of America Bulletin, 113: 1423-1442.

48. Miller, R. B., Paterson, S. R., Lebit, H., Alsleben, H., and Lunberg, C. 2006. Significance of composite lineations in the mid–deep crust: a case study from the North Cascades, Washington. Journal of Structural Geology 28: 302-322.

49. Miller, R. B, Paterson, S. R., Matzel, J. P. 2009. Plutonism at different crustal levels: Insights from the ~5 - 40 km (paleodepth) North Cascades crustal section, Washington. Geological Society of America Special Paper 456: 1-25.

50. Misch, P. 1966. Tectonic evolution of the northern Cascades of Washington State —a west-Cordilleran case history. In H. C. Gunning. Canadian Institute of Mining and Metallurgy, Vancouver, B.C., 1964, Spec. Vol. 8: 101-148.

51. Mitrovic, I. 2013. Evolution of the Coast Cascade Orogen by tectonic thickening and magmatic loading: The Cretaceous Breakenridge Complex, Southwestern British Columbia: M.Sc. thesis, Simon Fraser Univ., 133 p.

52. Monger, J.W.H. 1989, Geology of the Hope and Ashcroft map areas, British Columbia maps 41-1989 and 42-1989, Geological Survey of Canada.

53. Monger, J. W. H., van der Heyden, P., Journeay, J. M., Evenchick, C. A., and Mahoney, J. B. 1994. Jurassic-Cretaceous basins along the Canadian Coast belt: their bearing on pre-mid-Cretaceous sinistral displacements. Geol., 22: 175-178.

54. Monger, J. W. H., and Brown, E. H. 2015. Tectonic evolution of the southern Coast-Cascade Orogen, northwestern Washington and southwestern British Columbia. *In* The Geology of Washington and Beyond. *Edited by* E. Cheney. Washington Division of Geology and Earth Resources, Seattle: 101-130.

55. Monger, J. W. H., and Ross, C. A. 1971. Distribution of fusulinids in the western Canadian Cordillera. Canadian Journal of Earth Sciences, 8: 259-278.

56. Monger, J. W. H., Price, R. A., and Tempelman-Kluit, D. J. 1982. Tectonic accretion and the origin of the two major metamorphic and plutonic welts in the Canadian Cordillera. Geology, 10: 70-75.

57. Mooney, W. D., and Weaver, C. S. 1989. Regional crustal structure and tectonics of the Pacific coastal states: California, Oregon, and Washington. Geological Society of America Memoir 172: 129-161.

58. Moores, E. M., and Twiss, R. J. 1995. Tectonics. W.H. Freeman, New York.

59. Mustoe, G. E., and Leopold, E. B. 2014. Paleobotanical evidence for the post-Miocene uplift of the Cascade Range. Canadian Journal of Earth Sciences, 51: 809-824.

60. Paterson, S. R., and Miller, R. B. 1998. Magma emplacement during arc-perpendicular shortening: An example from the Cascade crystalline core, Washington. Tectonics, 17: 571-586.

61. Paterson, S. R., Miller, R. B., Alsleben, H., Whitney, D. L., Valley, P. M., and Hurlow, H. 2004. Driving mechanisms for >40 km of exhumation during contraction and extension in a continental arc, Cascades core, Wash. Tectonics. 23: TC3005.

62. Parrish, R. R. 1983. Cenozoic thermal evolution and tectonics of the Coast Mountains of British Columbia: Fission track dating, apparent uplift rates, and patterns of uplift. Tectonics 2: 601-631.

63. Ramos, V. A., Zapata, T., Cristallini, E., and Introcaso, A. 2004. The Andean thrust system--Latitudinal variations in structural styles and orogenic shortening. *In:* K. R. McClay, ed., AAPG Memoir 82: 30– 50.

64. Ragan, D. M. 1963. Emplacement of the Twin Sisters dunite, Washington. American Journal of Science, 261: 549-565.

65. Reamsbottom, S. B. 1974. Geology and metamorphism of the Mount Breakenridge area, Harrison Lake, British Columbia. Ph.D. thesis, Vancouver, University of British Columbia: 155p.,

66. Reiners, P. W., Ehlers, T. A., Garver, J. I., Mitchell, S. G., Montgomery, D. R., Vance, J. A., and Nicolescu, S. 2002. Late Miocene exhumation and uplift of the Washington Cascade Range. Geology, 30: 767-770.

67. Roddick, J. A. and Hutchinson, W. W. 1969. Northwestern part of Hope Map-area, British Columbia, (92H (west half)). Geological Survey of Canada Paper 69-1A: 29-38.

68. Shea, E. K., Miller, J. S., Miller, R. B., Bowring, S. A., Sullivan, K. M., 2016. Growth and maturation of a mid- to shallow-crustal intrusive complex, North Cascades, Washington. Geosphere, v. 12, no. 5: 1480-1516.

69. Schermer, E. R., Hoffnagle, E., Brown, E. H., Gehrels, G. E., and McClelland, W. C. 2018. U-Pb and Hf isotopic evidence for a paleo-arctic origin of terranes in northwestern Washington. Geosphere, v. 14 no. 2.

70. Schertl, H.-P., Schreyer, W., and Chopin, C. 1991. The pyrope-coesite rocks and their country rocks at Parigi, Dora Meira Massif, Western Alps: detailed petrography, mineral chemistry and PT-path. Cont. Min. Petrol., 108: 1-21.

71. Schmid, S. M., Pfiffner, O. A., Froitzheim, N., Schonborn, G., and Kissling, E. 1996. Geophysical-geological transect and tectonic evolution of the Swiss-Italian Alps. Tectonics, 15: 1036-1064.

72. Stowell, H., Bulman, G., Tinkham, D., and Zuluaga, C. 2011. Garnet growth during crustal thickening in the Cascades Crystalline Core, Washington, U.S.A. Journal of Metamorphic Geology, 29: 627-647.

73. Tabor, R. W., Zartman, R. E., and Frizelll, A. Jr. 1987. Possible tectonostratigraphic terranes in the North Cascades crystalline core, Wa.: Selected papers on the geology of Washington: Wa. Div. of Geol. and Earth Res Bulletin 77: 107-127.

74. Tabor, R. W., Haugerud, R. A., Hildreth, W., and Brown, E. H. 2003. Geologic Map of the Mt Baker 30- by 60-Minute Quadrangle, Washington. U.S. Geological Survey Miscellaneous Investigations Map I-2660, scale 1:100,000.

75. Tatsumi, Y. 2005. The subduction factory: how it operates in the evolving Earth. GSA Today, v. 15: 4-10.

76. Thompson, G. A., and Robinson, R. 1975. Gravity and magnetic investigation of the Twin Sisters dunite, northern Washington. Geological Society of America Bulletin, 86: 1413-1422.

77. Till, C. B., Grove, T. L., Withers, A. C. 2012. The beginnings of hydrous mantle wedge melting. Contributions to Mineralogy and Petrology v. 163: 669-688.

78. Umhoefer, P. J., Schiarriza, P., and Robinson, M. 2002. Relay Mountain Group, Tyaughton-Methow basin, southwest British Columbia: a major Middle Jurassic to Early Cretaceous overlap assemblage. Can. Jour. Earth Science, 39: 1143-1167.

79. Valley, P. M., Whitney, D. L., Paterson, S. R., Miller, R. B., and Alsleben, H. 2003. Metamorphism of the deepest exposed arc rocks in the Cretaceous to Paleogene Cascades belt, Washington: evidence for large-scale vertical motion in a continental arc. Journal of Metamorphic Geology, 21: 203-220.

80. Varsek, J.L. et al. 1993. Lithoprobe crustal reflection structure of the Southern Canadian Cordillera. Tectonics v. 12: 334-360.

81. Walker, N. W. and Brown, E. H. 1991. Is the southeast Coast Plutonic Complex the consequence of accretion of the Insular superterrane? Evidence from U-Pb zircon geochronology in the northern Washington Cascades. Geol. v.19: 714-717.

82. Wernicke, B., and Getty, S. R. 1997. Intracrustal subduction and gravity currents in the deep crust: Sm-Nd, Ar-Ar, and thermobarometric constraints from the Skagit Gneiss Complex, Washington. Geol. Soc. Am. Bull, 109: 1149-1166.

83. Wheeler, J. O., and McFeely, P. 1991. Tectonic assemblage map of the Canadian Cordillera and adjacent parts of the United States of America. Geological Survey of Canada Map 1712A, 1:2,000,000

84. Whitney, D. L., and McGroder, M. F. 1989. Cretaceous crustal section through the proposed Insular-Intermontane suture, North Cascades, Washington. Geology, 17: 555-558.